"This inspiring and informative book asks how we might construct material structures using biology. Its author, an architect and qualified synthetic biologist, is uniquely able to combine ambitious design and down-to-bench realism. Constructed around ideas – of life, of design, of fabrication – the book provides an ideal springboard to a biological architecture grounded, not in conceptual fantasies, but on what might really be achieved."

        Jamie A. Davies, *Professor of Experimental Anatomy, University of Edinburgh*

"*Living Construction* is a readable synthesis of important principles for the new field of biodesign, written by someone with graduate training in both architecture and synthetic biology. Dade-Robertson clearly knows the details but has the gift of extrapolating these into accurate yet broad generalities. While design and biology each has its disciplinary theories and practices, this book distills sound principles for their intersection in biodesign, offering a very useful contemporary map for practitioners in the arts and sciences. I especially admire his answer to his own question, 'where is the information in biological assembly?', for how it addresses multiple scales simultaneously. This is a key primer for all students in biodesign."

   Christina Cogdell, *Professor, University of California at Davis*

Living Construction

Modern biotechnologies give us unprecedented control of the fundamental building blocks of life. For designers, across a range of disciplines, emerging fields such as synthetic biology offer the promise of new sustainable materials and structures which may be grown, are self-assembling, self-healing and adaptable to change. While there is a thriving speculative discourse on the future of design in the age of biotechnology, there are few realized design applications.

This book, the first in the Bio Design series, acts as a bridge between design speculation and scientific reality and between contemporary design thinking, in areas such as architecture, product design and fashion design, and the traditional engineering approaches which currently dominate biotechnologies. Filled with real examples, *Living Construction* reveals how living cells construct and transform materials through methods of fabrication and assembly at multiple scales and how designers can utilize these processes.

Martyn Dade-Robertson is Professor of Emerging Technology at the School of Architecture, Planning and Landscape at Newcastle University and the Co-Director of the Hub for Biotechnology in the Built Environment. He holds degrees in architecture, architectural computing and synthetic biology.

Bio Design
*Series editor: Martyn Dade-Robertson,
Newcastle University, UK*

The Bio Design series offers the opportunity for designers from fields as diverse as architecture, fashion design and product design to present and explore designs and design research which use living systems as part of their production and operation. The series offers readers in depth project descriptions, analysis of processes and the intellectual contexts of Bio Design. Such explorations have not, until now, been made available in long form. The series also allows designers to explore the potentials and challenges of Bio Design as an emerging field of design and research.

The Bio Design book series will distinguish between key areas within the field, including, design fictions, biomimicry, bioinspired design and the use of biological materials and systems. While open to a range of voices the series will, as a collection, offer unified framework for thinking about Bio Design opening up this new rapidly growing field to a new generation of designers and researchers.

Living Construction
Martyn Dade-Robertson

Living Construction

Martyn Dade-Robertson

Taylor & Francis Group

LONDON AND NEW YORK

First published 2021
by Routledge
2 Park Square, Milton Park, Abingdon, Oxon OX14 4RN

and by Routledge
605 Third Avenue, New York, NY 10158

Routledge is an imprint of the Taylor & Francis Group, an informa business

© 2021 Martyn Dade-Robertson

The right of Martyn Dade-Robertson to be identified as author of this work has been asserted by him in accordance with sections 77 and 78 of the Copyright, Designs and Patents Act 1988.

All rights reserved. No part of this book may be reprinted or reproduced or utilised in any form or by any electronic, mechanical, or other means, now known or hereafter invented, including photocopying and recording, or in any information storage or retrieval system, without permission in writing from the publishers.

Trademark notice: Product or corporate names may be trademarks or registered trademarks, and are used only for identification and explanation without intent to infringe.

British Library Cataloguing-in-Publication Data
A catalogue record for this book is available from the British Library

Library of Congress Cataloging-in-Publication Data
Names: Dade-Robertson, Martyn, 1979– author.
Title: Living construction / Martyn Dade-Robertson.
Description: Abingdon, Oxon ; New York : Routledge, 2020. | Includes bibliographical references and index.
Identifiers: LCCN 2020010849 (print) | LCCN 2020010850 (ebook) | ISBN 9781138363014 (hardback) | ISBN 9781138363038 (paperback) | ISBN 9780429431807 (ebook)
Subjects: LCSH: Design and technology. | Synthetic biology.
Classification: LCC NK1520 .D35 2020 (print) | LCC NK1520 (ebook) | DDC 745.4—dc23
LC record available at https://lccn.loc.gov/2020010849
LC ebook record available at https://lccn.loc.gov/2020010850

ISBN: 978-1-138-36301-4 (hbk)
ISBN: 978-1-138-36303-8 (pbk)
ISBN: 978-0-429-43180-7 (ebk)

Typeset in Calvert
by Apex CoVantage, LLC

Contents

Series editor's preface   ix
Ethics and society   xi
Acknowledgements   xiii

1 Introduction   xvi
    Growing buildings   1
    Living architecture   1
    Synthetic biology   2
    A new approach?   3
    Biological design   3
    Scope of the book   4
    A note on the living   6
    Structure of the book   6
    Ready?   8

2 The designs of the natural   12
    Blurring between the natural and the artificial   13
    Biology as machine   14
    The sciences of the artificial   15
    Inside and outside   16
    Artificializing the natural in synthetic biology   17
    Forcible constraint v. nature's own agencies   21
    Beyond the machine   23
    Conclusion   25

3 The logic of living assembly   30
    Biological construction: between assembly and fabrication   31
    Logics of assembly in alien technology   33

Chemistry: bounding, concentration and the patterning of forces	35
Inside a cell: patterning of parts and the constraint of forces	36
Multicellular assembly: patterning of environment	42
Emergence: information for free	47
Self-assembly and external control	50
Beyond the 'self'	52
Conclusion	52
4 Fabrication in the living	**56**
From assembly to fabrication	57
Top down v. bottom up	60
In vivo and *in vitro* biological fabrication	61
Bottom-up causality: Thinking Soils	67
Top-down causality: bio-induced mineral crystals	73
The challenge of decomposability	76
Diagram 1: domains of information	78
Diagram 2: phase spaces and landscapes of change	79
Composing the music of life: creodic design	86
Conclusion	89
5 Conclusion: the craft of living construction	**94**
The end of the beginning	95
Crafting biology	97
A new agricultural revolution	100
Index	102

Series editor's preface

The emergence of biotechnologies and their integration into human-centred contexts of use has led to a new design paradigm known as Bio Design. Bio Design, bio-design or biodesign, depending on the source, has a wide range of definitions. Early examples of Bio Design are mainly focused on the design of medical devices and methods to support tissue engineering. Bio Design has also been used alongside fields such as synthetic biology. This series, however, uses a broader definition of Bio Design which includes the integration of biological processes across a range of creative design fields including, but not limited to, architecture, fashion and apparel design, product design and interaction design.

Too often, real applications and technologies are confused with grandiose pronouncements of technological revolution based on science fictions built on limited or no real engagement with messy biological reality. The series will allow for deeper explorations of Bio Design projects and theories. The books will necessitate critical and reflective descriptions of the Bio Design field focused on the underlying processes, methods and theories.

In defining the territory of this book and the series it initiates, we are seeking a happy middle ground between design speculation and grounded experiment, between critical thinking and creative naivety, between formal elegance and radical complexity and to provide ways of thinking which will lead to critical ways of making.

Martyn Dade-Robertson
Newcastle University
February 2020

Ethics and society

Any conversation on designing with biology necessitates reflection about ethics. Novel biotechnology and bioengineering applications have the potential to provide enormous benefits for society. However, the relationship between human activities and the current climate emergency due to global warming remind us that technologies used to solve one set of problems are capable of creating many others. Beyond environmental impacts, there is also the challenge of considering diverse political and social values in the development of bioengineered technologies and materials. As a result, we wanted to establish an ethical position in developing this series.

- We will not seek to publish material which could be applied directly to the development of weapons or deliberate causes of harm.
- Where experiments or processes are introduced which may be harmful to the individuals conducting them or the environment, we will make these risks clear, especially given audiences who may be unfamiliar with the techniques and technologies being described.
- Authors in the series will be required to confirm (where applicable) that appropriate risk assessment, ethics review, informed consent and animal welfare protocols have been met, in compliance with local institutional and governmental regulations.

While we make every effort to anticipate risks in our research, the unintended consequences of technology are harder to predict. However, in formulating this series, we believe that Bio Design is at its best when it is a reflective and inclusive practice where ethical and responsible principles are embedded. The books and editorial guidance in this series prioritize this, while accepting diversity of opinion and position.

Martyn Dade-Robertson
Carmen McLeod
Newcastle University
February 2020

Acknowledgements

When I first pitched the idea of the Bio Design book series to Routledge, it was with some selfish intent. The idea of writing a full monograph of 80,000 (or more) words filled me with horror and flashbacks to my PhD. But writing 5000- to 8000-word journal papers didn't give me the space to properly explore my ideas. Twenty-five to thirty thousand words sounded like a sweet spot. How little I knew. This book turns out to have been one of the most difficult texts I have written (almost harder than my PhD). What started as a review, trying to sum up my ideas on biological construction and feature some of the work in my lab, became a wrestling match between my newfound interest in synthetic biology and my background as a designer with a humanities bent (I confess to my students that, at architecture school, I used to regularly fail my architectural technology exams). I've written more than 80,000 words to get to this point and would (although not happily) write the whole thing again. This was a battle between the left and right sides of my brain, and it isn't clear who won. What is clear is that I couldn't have gotten this far without help.

Newcastle University has provided me, both as an undergraduate student and as an academic, with a nurturing environment to develop. I've often not taken a conventional route, but when I asked John Pendlebury and Graham Farmer, the Heads of School and Department at the time, for a sabbatical to pursue an MSc degree in synthetic biology, it is still a mystery to me why they said yes. The experience, however, has been transformative, and I am forever grateful for the opportunity they gave me and the continued support of colleagues, especially Adam Sharr.

During my studies in this area, I want thank the staff and students on the synthetic biology course, especially Anil Wipat, Natalio Krasnogor and Wendy Smith and my co-students Zelda Mendelowitz, James Skelton and Stephen Bradley. Many of the ideas here have been developed in conversation with them.

Some of the projects I have described here are supported by research funding, including the pressure-sensing bacteria work, which has been funded by the Engineering and Physical Sciences Research Council through two grants: Computational Colloids (EP/N005791/1) and Thinking Soils (EP/R003629/1). These results described and the concepts developed are a big team effort, and special thanks should go to the team, including Anil Wipat, Helen Mitrani, Meng Zhang, Aurelie Guyet, Beate Christgen, Polly Moreland, Javier Rodriguez Corral, Jennifer Wright and Jamie Haystead. This work has also been supported by pilot projects resourced by both Newcastle and Northumbria University, and the Hub for Biotechnology in the Built Environment is funded by Research England. Thank you to Katia Zolotovsky and Merav Gazit for their permissions for Figure 4.5. Their Guided Growth project cited in Chapter 4 was funded by the National Science Foundation Division of Materials Research (NSF DMR) under the grant #1508072 named "Material and morphometric control of bacterial cellulose via genetic engineering post-processing and 3D printed molding".

I'd also like to thank Lynn Rothschild (NASA Ames) and Jamie Davies (Edinburgh University). Both are scientists I have huge respect for, and this book has been enriched by conversations I have had with them. They have taken my (sometimes sketchy) ideas seriously. Thank you also to Marcos Cruz and Philip Beasley for many fascinating conversations and opening up new opportunities and networks.

While we like to talk about 'research-led teaching' in academia, less focus is given to 'teaching-led research'. Many of the ideas in the book have been developed in design studios and supervisions. I would need to radically increase the word limit of this book to cite them all, but special mention should be given to Carolina Ramirez-Figueroa, who was my first PhD student in Bio Design and a trailblazer at a time when we had few resources to do this sort of work, and Luis Hernan, who developed our first lab and helped build the group in challenging circumstances. The future of this field lies in the hands of PhD students like those in my group, and I would like to mention Thora Arnadottir, Dilan Ozkan, Sunbin Lee, Emily Birch and Aileen Hoenerloh. Of the dozens of undergraduate and postgraduate students I have supervised, I want to give a special mention to my master of architecture students, whose work on biomineralization is featured in Chapter 4: John Beattie, Alexander Lyon, Markus Ryden and Malcolm Welford.

Finally, to my family. My parents, whose love and support has underpinned this book and everything else I have written. To my children, who are my most advanced living construction experiments so far. And to my wife, Meng. The love of my life, my collaborator and my inspiration.

Introduction
Chapter 1

GROWING BUILDINGS

In 2010, an article in the *Mail Online* presented the work of a group of students who, the headline claimed, had developed a "glue made from genetically-modified bacteria that can knit cracks in concrete back together" (Firth, 2010). The undergraduate team, composed of bioscientists and computer scientists as well as engineers, were pioneers in a new field called *synthetic biology*. They had won a gold medal in the International Competition for Genetically Engineered Machines (iGEM) by designing and building a living system composed of bacteria cells which were capable of swimming deep into microscopic concrete cracks and then, using a chemical signalling process known as quorum sensing, commencing a process of biomineralization. In addition, the cells would excrete a biological glue and become filamentous, growing into long fibrous strands. Through this process, cracks in reinforced concrete, which otherwise would have caused water ingress to rust the steel tension bars, would be intelligently filled.

I read a version of this article on the Newcastle University news feed and was intrigued. The work had been conducted in a lab less than five minutes' walk from my office, and my first reaction was to ask why no one from my School (of Architecture, Planning and Landscape) was involved.

A few months later, I found myself at a conference (where I delivered a paper on a very different topic) gossiping with delegates about the iGEM project, and the science fiction fan in me began to project possible scenarios. If you could repair concrete with bacteria, why not propose a world in which buildings would be grown using the same approach?

LIVING ARCHITECTURE

Nine years later, after completing a master's degree in synthetic biology, running a small, but rapidly growing, microbiology lab and conducting a first tentative set of experiments, I began to seek out a community of like-minded researchers. Whilst this area of research was lacking an established scientific community, I discovered a growing collection of speculative projects and practitioners in the field of *living architectures*. Communicating as much through TED talks and blog posts as research journals, a new generation of architectural designers had begun to speculate on a future in which our buildings would be grown from living cells, self-assembled and responsive to their environment, capable of adapting to change, self-healing and even capable of reproduction. However, while these visions were often inspiring, they can become disconnected from scientific and biological reality. They offer speculation on the *what?*, sometimes the *why?*, but rarely the *how?* Too often, dreams of living architectures are associated with notions of biological form or computational and generative architecture. As Cogdell points out in her critique of living architectures, terms such as genetic or morphogenesis, borrowed from biological textbooks, often become confused and misused (2018). At worst, living architectures offer empty renderings based on what Michel Hensel describes as the "superficial biomorphic formal repertoire" (2006). Speculation in itself is not a problem. Design is inevitably a speculative act. Architecture has thrived on cultural and technical speculation, but it is a tough reality that few designers have access to the skills or resources required to carry out the sorts of experiments necessary to move beyond speculation.

SYNTHETIC BIOLOGY

If the answers to the *how* are not to be found in design speculation, then perhaps they lie in synthetic biology. Synthetic biology (at least in its most contemporary formulation) uses the tools of molecular biology together with the principles and language of traditional engineering. Biological systems are described in terms of interchangeable parts and hierarchies of abstraction as systems are built from simple components to complex wholes (Benner and Sismour, 2005). In the narrative offered by synthetic biology, biological systems can be redesigned along more logical and elegant lines and be made predictable and precise.

Practitioners in the field of synthetic biology benefit from access to wet labs and substantial research resources but are, in other ways, also profoundly limited. The technologies offered by synthetic biology are modest and the approaches are gene centric, rendering living cells as wetware machines running on a software code of deoxyribonucleic acid (DNA). Time spent at a lab bench reveals that this software code can (after many hours of work) be edited to, for example, engineer a bacteria to produce a new protein or to respond to simple environmental cues, but there are no genes to assemble a few billion bacteria into a cuboid to produce a brick, let alone a building.

A NEW APPROACH?
For me, as an architect, an irony of the study of synthetic biology is that at the same time as the discipline is looking to frame biology through methods derived from industrial engineering, designers in my field are often using biology as a way of critiquing traditional engineering approaches to design. Bioinspired and biomimetic design often cites biology as offering a fundamentally different logic of material construction. This discourse can be found, for example, in discussions of material ecology and material computation in the work of Oxman (2010) and Menges (2012) and of self-assembly in the work of Tibbits (2017) and in the experimental architectures of Armstrong (2015). In referring to nature's materials and structures, they cite the seamless way in which multiscale biological materials are formed into complex multiscale structures. By seeking inspiration from biology, designers frequently make references to the complexity of biological systems, their interconnectedness with their environment and ecologies and their irreducibility to parts and hierarchies.

BIOLOGICAL DESIGN
The absence of a firm knowledge base for their work doesn't stop a steady and growing stream of design students approaching me and my colleagues wanting to know how they might make their design speculations real. These students are operating in a new field of Bio Design, a context which encompasses architecture, fashion design, interaction design and product design, among others. These new bio designers are not satisfied with speculation alone. They want to make. They see the lab as a potential extension of their studios and workshops. However, the incredible complexity, not just of biology but of the way in which biological knowledge is codified, is daunting. To the students who want to grow a building canopy with bacteria, produce cellulose or grow a meat house, my answer is invariably: 'it's a bit more complicated than that'.

Most students, horrified by the pile of scientific papers we give them, and the realization that success in my group is often measured in material samples of a few micrometres, retreat back to Gaudi-like renderings of future living buildings grown through as yet undefined technologies. Some, however, persevere. These brave souls persist, cementing ammonia-smelling sand columns with *Sporosarcina pasteurii*, growing cellulose membranes with kombucha tea and growing blocks of mycelium (the roots of fungus) on coffee grounds. They may not yet produce buildings, but they get to observe miraculous transformations of matter and exquisite complexity through the lenses of a microscope. Those who stay long enough begin to develop their expertise and manual skills in experimental biology. They also develop an intuition which goes beyond their scientific knowledge to include experimental practices akin to a craft.

To start on this road, however, is challenging. The confusing hype around fields such as synthetic biology and living architecture and the gap between biological reality and speculation is substantial. This book is my attempt to fill this gap. It reflects the way that I have wrestled with the challenges of Bio Design, attempting to translate biological knowledge from fields like synthetic biology into terms and models that, as a design academic, I (and hopefully others) can understand. It is also an attempt to reconcile the richness and complexity of biology with a need to develop formal methods in design by looking towards frameworks that can be taught to others and guide our experiments.

SCOPE OF THE BOOK

Before elaborating on what this book is, it is worth saying what it is not. First, this is not a biological textbook. The way I describe biological processes is simplified, and some of my terms and schematic representations would be alien to biologists. A recurring theme in this book is the relation between the map and the territory. Even simple biological systems such as single-cell bacteria are unimaginably complex, so to navigate the territory of living cells, we need a vast array of maps depending on our navigational needs. If we were to extend the analogy of navigation, then synthetic biologists often make use of maps which are more like subway maps – they tell you little about the territory of the city but give you enough information to travel from place to place along specific routes. Pick up a cell biology textbook, however, and you find an array of detailed representations more like an atlas, which describe the geometry of individual buildings (e.g. the structure of proteins) or topographies of whole cities (e.g. cells). There is no shortcut to understanding the territory of living cells, and there are plenty of good textbooks which act as atlases (I have often used Alberts *et al.*, 2014, for example). This book uses an alternative cartographic language which, through text and diagrams, attempts to describe biological principles through the lens of design, making reference to pattern and form. The book will not, therefore, provide a deep understanding of biological processes but rather a useful set of schematics.

The book will also not provide recipes for biological materials or structures. There are, however, a growing number of DIY biology setups supported by initiatives such as MIT's Bio Summit, competitions such as iGEM (no date) and the Biodesign Challenge (no date) and methods are also shared in countless blog posts and community resources such as Materiability.com (no date). There are also a growing number of synthetic biology how-to guides including general introductions (Davies, 2018) and practical primers (Baldwin *et al.*, 2009; Kuldell *et al.*, 2015). While I will cite a number of practical examples throughout the book, these are intended simply to ground a set of broader concepts and principles.

What this book will address is the question: How do we construct material structures using biology? With a background in architecture, my interest is in constructing human-scale structures. This book, however, is much broader than that. We have been using nature's bounty of geological and biological materials in construction for as long as we have been creating buildings. So there is nothing inherently new about this question. We have recognized the qualities of wood as a construction material, harvesting and cutting it into usable forms. In turn, the functional properties of timber have shaped the products made from it, defining the structural spans of timber beams, for example. Biological materials, including but not limited to wood, have proven to be versatile, but our needs have also become more demanding.

With a seemingly unlimited supply of energy and matter, we can demand precision from our materials by applying heat and pressure or changing their chemical structure into new forms. Plastic, after all, is actually a biological material made, most often, from hydrocarbons from oil, but, in its highly processed state, it has become something synthetic, gaining new properties which make it highly versatile. Energy and matter, however, are not inexhaustible, at least in the way we currently use them. Plastics have proven to be highly problematic, requiring large amounts of energy in their production (Johnson, 2015), and moving away from their original biological source has meant that the resulting material is resilient in ways which are incompatible with our ecosystem.

This book starts from the premise that we might begin to harness a much greater range of biological materials, many of which don't yet exist, by thinking of materials not as matter to be harvested after the death of the organism which created it but by directing the process of material making while the organism is alive. To do this, however, we need to think quite differently about our processes of manufacture and our ideas of assembly and fabrication.

A NOTE ON THE LIVING

The book title, *Living Construction*, necessitates a definition of *life*. Definitions of life are varied, especially in design. 'Living architecture', for example, through initiatives such as the Living Architecture Systems Group (LASG) (Beesley, Hastings and Bonnemaison, 2019) and Built to Grow (Imhof and Gruber, 2015), includes technologies and materials that are life-like and do not necessarily involve biology but can include responsive systems driven by computation. Similarly, work on living technologies can include protocells (Armstrong and Spiller, 2011) which are *life-like* but are pre-biotic. Computer systems can also create *artificial life* where their software demonstrates growth, self-replication, responsiveness to the environment, metabolism and the capability of evolution. Definitions also change as scientific knowledge changes (Luisi, 2006), but for the purposes of this book, life is defined by biological cells. In this book, I use the term *biological systems* to describe both single-celled organisms and multicell organisms. Because, in synthetic biology, much of the pioneering work has involved micro-organisms, especially bacteria, and because my own background is in microbial synthetic biology, bacteria will be referenced more than other forms of life.

In all cases of biotic life, the cell is a fundamental unit of matter. Living cells are a material in their own right and are also the cause of extracellular (outside the cell) material construction, and this book is interested in both.

STRUCTURE OF THE BOOK

If digital technology dominated our discussions of design at the end of the 20th century, it is sometimes suggested that the 21st century will be dominated by biological technologies and that a new technological revolution will lead to *the end of the artificial*. The terms we use play a significant role in defining the practices and outcomes of fields such as synthetic biology, and Chapter 2 will offer a critique of this view using, as a framing device, Herbert Simon's book *The Sciences of the Artificial* (1996). The chapter will investigate the historical and philosophical distinctions between natural and artificial and how they may be applied to design. It will show that the distinction has never been entirely clear, but, rather, there is an evolving relationship between the wild and the cultured in the way we shape and perceive the material world. This will set some of the key ideas and parameters for the rest of the book, critiquing our current understating of synthetic biology and introducing ideas of the *inside* and *outside* of designed systems.

Chapters 3 and 4 will begin to describe the processes of *assembly* and *fabrication* in biology.

In Chapter 3, we will ask "Where is the information?" and I will argue that our ways of designing biological systems are dependent upon finding and altering sources of information rather than interacting directly with the parts themselves. However, I will also question the idea of self-assembly – questioning the idea of a discrete *self* and beginning to identify sources of information which cross scales.

Chapter 4 will then go on to investigate the idea of fabrication in biological systems. As a result of our methods for analysing biological systems and our structuring of knowledge in science, we tend to see biological systems in terms of stratified scales. Different disciplines of science are built around apparatus designed to understand scales of assembly from molecules to whole galaxies. This stratification leads, in discussions in Bio Design, to an argument as to whether the most appropriate design methods are top down or bottom up. However, as I will show, many of the effects we see in lab experimentation for the biological fabrication of materials show influences across scales. These multiscale *causalities*, I will argue, require us to think beyond the stratified multiscale approach we currently take.

To account for these causalities, I suggest in Chapter 4 the use of Waddington's theory of *creodes* or the *necessary path* and epigenetic landscapes in developmental biology. This model, I will propose, requires new ways of thinking about biological fabrication in terms of *pathways of change* rather than *hierarchies and parts*.

READY?
At the conference where I discussed the iGEM project and my dreams of growing buildings, I was faced with a combination of amusement and scepticism. The other delegates generally subscribed to what I later came to know as the technology readiness level (TRL) view of emerging technology and its application. The TRL view (see Figure 1.1), initially developed to grade the

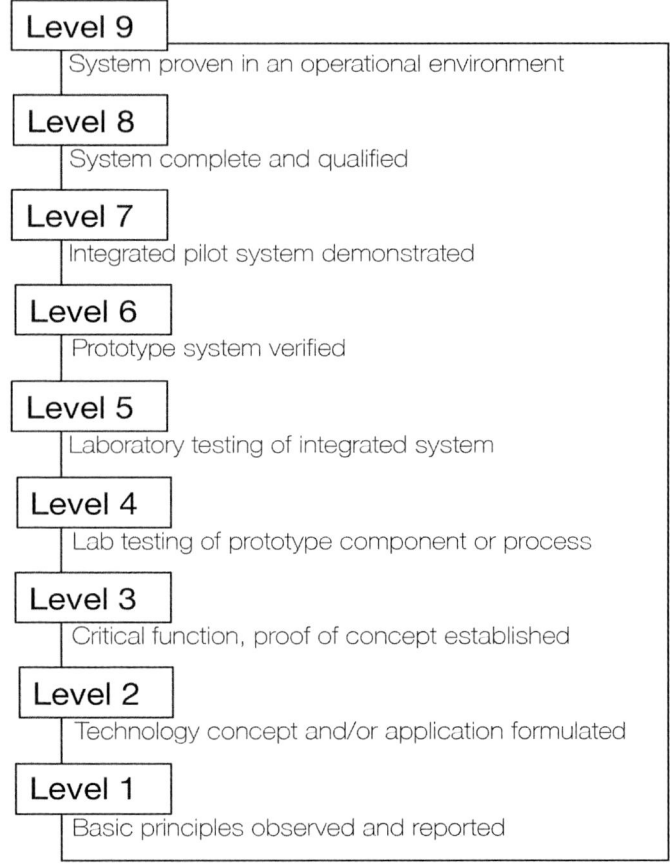

Figure 1.1 Technology readiness levels chart showing levels of maturity for new technologies, originally devised by NASA.

mission readiness of R&D projects at NASA, describes the status of technology in terms of nine levels, from fundamental or basic technology research (Level 1) through to demonstrators (Level 6) and mission-ready operational systems in real environments (Level 9). Variations of these descriptors are at the heart of many research funding organizations. Not surprisingly, architectural technologists often look towards the top of the TRL chart – preferring to use technologies which can be immediately deployed and tested in real environments. That is, after all, what we are normally attempting to do as architectural researchers.

Six years later I wrote an article in which I noted that:

> when I embarked on a postgraduate degree in Synthetic Biology, I would have described the state of the art in Synthetic Biology with reference to computing in the 1960s and '70s which was then poised to have a transformative effect on society. . . . If I were to draw the same parallel now, however, I would describe Synthetic Biology as being at its Babbage stage – with foundational concepts forming but with few practical demonstrators yet and fundamental questions remaining.
> (Dade-Robertson, 2016: 6)

This, for me, is an important point, and this book is a first step towards a new relationship between biology and the constructed world, one which, as Armstrong points out (2015), will see us as co-creators of new material structures and architectures. We need to recognize that our research may reside for some time at TRL levels 1, 2 or 3. Our collaboration with the living is one in which we are only just beginning to understand the language of our collaborators.

REFERENCES

Alberts, B. *et al.* (2014) *The Molecular Biology of the Cell.* 5th edn. New York: Garland Science.

Armstrong, R. (2015) *Vibrant Architecture: Matter as Codesigner of Living Structures.* Warsaw: De Gryter Open.

Armstrong, R. and Spiller, N. (2011) 'A Manifesto for Protocell Architecture Against Architectural Formalism', *Architectural Design*, 210, pp. 24–25.

Baldwin, G. *et al.* (2009) *Synthetic Biology: A Primer.* London: Imperial College Press.

Beesley, P., Hastings, S. and Bonnemaison, S. (eds) (2019) *Living Architecture Systems Group: White Papers.* Waterloo: Riverside Architectural Press.

Benner, S. A. and Sismour, A. M. (2005) 'Synthetic Biology', *Nature Reviews: Genetics*, 6(7), pp. 533–543. https://doi.org/10.1038/nrg1637.

Biodesign Challenge (no date) *Biodesign Challenge*. Available at: https://biodesignchallenge.org (Accessed: 7 February 2019).

Cogdell, C. (2018) *Towards a Living Architecture: Complexism and Biology in Generative Design*. Minneapolis: University of Minnesota Press.

Dade-Robertson, M. (2016) 'Building Science: Synthetic Biology and Emerging Technologies in Architectural Research', *Architectural Research Quarterly*, 20(1), pp. 5–8.

Davies, J. A. (2018) *Synthetic Biology: A Very Short Introduction*. Oxford: Oxford University Press.

Firth, N. (2010) 'The Glue Made from Genetically-Modified Bacteria that Can Knit Cracks in Concrete Back Together', *Mail Online*. Available at: www.dailymail.co.uk/sciencetech/article-1331281/BacillaFilla-glue-bacteria-knit-cracks-concrete-together.html (Accessed: 1 November 2019).

Hensel, M. (2006) '(Synthetic) Life Architectures: Ramifications and Potentials of a Literal Biological Paradigm for Architectural Design', *Architectural Design*, 76(2), pp. 18–25. https://doi.org/10.1002/ad.236.

IGEM (no date) *International Genetically Engineered Machine Competition*. Available at: https://igem.org/Main_Page (Accessed: 31 October 2019).

Imhof, B. and Gruber, P. (eds) (2015) *Built to Grow: Blending Architecture and Biology*. Basel: Birkhauser.

Johnson, A. (2015) 'Recycling Energy: An Exploration of Recycling and Embodied Energy', *Penn Sustainability Review*, 1(6), p. 3.

Kuldell, N. *et al.* (2015) *BioBuilder: Synthetic Biology in the Lab*. Sebastopol: O'Reilly.

Luisi, P. L. (2006) *The Emergence of Life: From Chemical Origins to Synthetic Biology*. Cambridge: Cambridge University Press.

Materiability (no date) *Materiability*. Available at: http://materiability.com/ (Accessed: 19 October 2019).

Menges, A. (2012) 'Material Computation', *Architectural Design*, 82(2), pp. 14–21.

Oxman, N. (2010) 'Structuring Materiality: Design Fabrication of Heterogeneous Materials', *Architectural Design*, 80(4), pp. 78–85.

Simon, H. A. (1996) *The Sciences of the Artificial*. Cambridge, MA: MIT Press.

Tibbits, S. (2017) *Self-Assembly Lab: Experiments in Programming Matter*. Abingdon: Routledge.

The designs of the natural
Chapter 2

BLURRING BETWEEN THE NATURAL AND THE ARTIFICIAL

In the inaugural edition of the *Journal of Design and Science* (*JoDS*), Joi Ito made reference to the "End of the Artificial", suggesting that:

> Unlike the past where there was a clearer separation between those things that represented the artificial and those that represented the organic, the cultural and the natural, it appears that nature and the artificial are merging.
>
> (Ito, 2016)

In his own *JoDS* article, Danny Hillis suggests that "We can no longer see ourselves as separate from the natural world or our technology, but as a part of them, integrated, co-dependent, and entangled" (2016).

Such claims of an entanglement between the natural and the artificial are common among a new generation of biologically aware designers. In an era where the entire genome of a cell can be printed out and the cell rebooted to grow and reproduce 'naturally' (Gibson *et al.*, 2010), the line between 'natural' biology and 'artificial' technology seems to have blurred. Furthermore, acknowledgement of the radical changes that our technologies are making to our environment has highlighted that, for too long, we have seen our bubbles of technologized civilization as entirely separate from the natural world. High time, then, that this relationship be reframed.

It is difficult, however, to achieve a design practice based on a blurred distinction between the natural and artificial. Are we naturalizing the artificial or artificializing the natural? We have, surely, always been entangled with nature, even if we have been unaware of it or have chosen to ignore it.

In this chapter, I want to look at the distinctions we make between the natural and the artificial and assess the implications of these distinctions for design and for my proposed approach to biological construction. Specifically, I want to take on the machine metaphor which artificializes the natural. Following from the machine metaphor, I will argue, are design frameworks in fields such as synthetic biology which frame the design of biological system as akin to the design of traditionally engineered systems. This framework, while pervasive, is also problematic and contradicted by both the practices and products of Bio Design.

BIOLOGY AS MACHINE

The distinction between the natural and artificial and our definitions of both lie at the heart of Bio Design as a past, present and future practice. My own quest has been to develop ways of making cells make 'stuff'. More accurately, to use living bacteria cells as a medium for manufacturing materials. Connected to our idea of a biological medium are the tools and practices we use to shape the outcomes of our design intention. Give a student a plank of wood, and it may also be appropriate to provide them with saws and sandpaper and a manual on carpentry. Give the student an agar plate of growing bacteria, and the tools, both conceptual and literal, become more difficult to define. We have access to the paraphernalia of the biology lab, but many of these tools and their methods of use are designed for analysis and observation rather than creative synthesis.

In explanations offered to students, I often do what many have done before me: I invoke a machine metaphor. I liken the bacteria to robots. It is just, I argue, that we are working at a different scale and we are using wetware rather than software to control the way materials are created and distributed. The analogy serves a purpose. Robots have become fashionable in architectural design and are well understood by my students as being an advanced fabrication method. Robot arms, for example, are capable, with the right end effectors and configuration, of shaping materials through extrusion or carving. Bacteria can also extrude materials or create enzymes which will dissolve materials. Once set up, a robot will complete its task independently; it can, therefore, appear autonomous. So too can bacterial cells, as their actions appear independent of the observer. Robots are not, in the main, controlled directly but rather though programmed software. Bacteria have their own control methods, inputs, outputs and wetware controlled, it is sometimes suggested, through the code of DNA. By using this analogy, I am attempting to artificialize the natural and provide a design framework with parameters my students will understand.

The robot analogy, however, is limited and problematic. There is an inevitable raising of eyebrows from my students when I use the machine analogy to describe a living thing – even a 'simple' single-celled organism. Instinctively we feel that there is something different about life.

If my students are sceptical about the machine analogy in theory, they wholeheartedly reject it after a few days at the laboratory bench. Bacterial cells work at length scales which, most of the time, are not visible to the naked eye. However, in a single agar plate experiment, we can be dealing with more cells than there are humans on earth. Neither of these scales is perceivable or easily imaginable. Bacterial cells are also not nearly as programmable as robot arms, and they have much greater autonomy and range of behaviour. They might live or die, and they are sensitive to even small changes to their environment. On one day, they will grow happily; on another, they will decline radically for reasons we can't easily explain. If our robot arm were as temperamental, we would send it back to the manufacturer. Cells are not machines; they have a different 'nature'.

Despite this, the machine analogy, when describing biological systems, is pervasive. When considering any new medium, it is common to seek recourse to an earlier form. Cars had a 'horseless carriage' phase before we understood them as a new form of transport, and digital media tends to 'remediate' existing media forms through simulation, for example, by using paper-based metaphors in graphical user interfaces for personal computers (Dade-Robertson, 2011). The machine metaphor may not, therefore, be easy to shake off. "Each age," suggests Bensaude-Vincent, "tends to interpret nature through models derived from one of its most advanced technologies" (Bensaude-Vincent and Newman, 2007: 293). In this tradition, biological systems have consistently been understood through machine metaphors. Trace the history of biology and you find a vision of the natural world defined by the state-of-the art technologies of their time from the 17th-century clock of finely tuned parts through to cells being likened to computer-like information processors composed of feedback inputs, outputs and interfaces (Nicholson, 2013).

THE SCIENCES OF THE ARTIFICIAL

The central idea of the machine metaphor in biology is that natural objects can be framed in the same way as artificial objects. Definitions of natural and artificial are slippery and imperfect, but it is important to try to make some distinctions. To attempt this, I am going to use as a framing device Herbert Simon's *The Sciences of the Artificial* (1996) by asking: What happens when we make the simple exchange of terms by swapping out *sciences* for *designs* and the term *artificial* for *natural*?

The Sciences of the Artificial is now considered a classic and one of the founding texts of design science (Huppatz, 2015). The book brings together computer science, economics, political science and psychology (and more besides) through the common theme of designing artificial systems. Simon distinguishes between science, which is concerned with *analysis*, and engineering, which is concerned with *synthesis*. The designer is "concerned with how things ought to be" (1996: 4). Design, he argues, "distinguishes the professions from the sciences" (Simon, 1996: 111).

Simon's thesis emerged in response to the post-war culture in US universities which sought greater academic credibility by extending degree programs in natural sciences and closing professional programs considered "intellectually soft, intuitive, informal and cookbooky" (1996: 114). Simon's thesis addresses this problem by proposing programmes of research and teaching which are concerned with making design theory *explicit and precise* through a rational framework for professional decision-making. In other words, Simon was attempting to bring scientific rigour to the design of human systems and artefacts. Since the publication of the first edition in 1969, *The Science of the Artificial* has been widely critiqued and alternative theories of design proposed. These criticisms have tended to focus on the positivistic nature of his model, and alternative views of human design and problem solving have been proposed by, for example, Rittel (Protzen and Harris, 2010), who introduces the idea of wicked problems for which there are no optimum solutions, and Schon (1984), who emphasizes the role of improvisation, professional intuition and reflection in action in human practice. These critiques are well rehearsed, and I will not revisit them here, but I do want to cast Simon's text in a different light.

INSIDE AND OUTSIDE

In Simon's thesis, the term 'natural' is incompatible with 'design' unless we adopt a creationist viewpoint, which I don't intend to do. Simon, therefore, makes a clear distinction between things which are natural and subject to the observations and theories of natural science and things which are artificial and designed. Simon suggests a number of "boundaries" for the study of the artificial:

1   Artificial things are synthesised (though not always or usually with full forethought) by human beings.

2   Artificial things may imitate appearances in natural things while lacking in one or many respects the reality of the latter.
3   Artificial things can be characterised in terms of functions, goals and adaptations.
4   Artificial things are often discussed, particularly when they are being designed, in terms of imperatives as well as descriptives.

(Simon, 1996: 26)

Simon defines an *outside* 'natural' environment which acts as a mould which shapes the *inside* of 'artificial' things. Artificial things are thus defined by a relationship between an 'inner' and 'outer' world. He uses the example of commonly understood machines. A sundial, for example, will only work when the sun is shining and through its orientation and the organization of its gnomon and dial plate. A clock for use at sea without the fixed position required for a sundial uses a mechanism which is balanced to counter the motions of the waves. The natural is thus described in terms of the elemental forces of the sea and sun, and the artificial is clearly declared as the internal human-made mechanisms of the clock. Design is concerned with the inside of the machine and the interface between this inner mechanical and outside natural world. Simon also introduces the idea of homeostasis, using the example of the ship chronometer, which achieves its function through quasi-independence from its environment. The artificial inside is in this case insulated from the natural outside.

ARTIFICIALIZING THE NATURAL IN SYNTHETIC BIOLOGY

Natural should not, in Simon's description, be conflated with biological. The distinction between artificial and natural is partly about our terms of reference.

The Sciences of the Artificial was first published only 16 years after the structure of DNA was published (Watson and Crick, 1953) and well before many of the advances in biotechnology we see today. However, Simon's way of approaching design seems all the more relevant when we consider the emerging field of synthetic biology. I use the term 'emerging' advisedly since, depending on your perspective, synthetic biology is either a few years old (Kitney, 2012) or based on a tradition which has been around since the early part of the 20th century (Campos, 2009).

A contemporary idea of synthetic biology is coalescing around the definition of a discipline "to design and engineer biologically based parts, novel devices and systems as well as redesigning existing, natural biological systems" (Kitney, 2012: 6). This type of synthetic biology is based on three guiding principles:

1. To control nature, making living organisms to serve human needs and goals;
2. To frame the work of the synthetic biology practitioners as the creating of life through the "dissolution of the distinction between artificial and natural creations" (Kingsland, 1991) (7) and
3. To reimagine biology as a discipline concerned with synthesis rather than analysis (Pauly, 1987).

Underlying 20th-century approaches to synthetic biology has also been a call for precision. Before the molecular revolution in synthetic biology, initiated in the 1940s, biological science was seeking the rigour provided by the reductionism of physics or chemistry, starting with scientific breeding programmes (Kay, 1991). Through these endeavours to control and ultimately construct life, the machine metaphor has also persisted.

Synthetic biology, in its modern form, is often associated with the genetic manipulation of organisms and attempts to systematize biological knowledge and to standardize descriptions of gene-level biological processes such that they can be engineered to create new systems (relatively) easily and reliably (Endy, 2005). This type of synthetic biology is influenced by computing science and electrical engineering and conceptualizes individual genes as parts which can be organized to create genetic circuits capable of responding to a range of inputs by synthesizing proteins and other molecules which have useful applications.

A cell only makes the proteins which it needs, so genes can be turned on and off or up and down depending on what proteins are needed at different times. Gene regulation is a highly complex topic, but, for a synthetic biologist, the mechanisms of gene regulation can be treated like switches in a circuit. For example, if we can identify genes which are only produced in the presence of a specific chemical trigger, then we can make a gene 'switch' which is turned on in response to the chemical and off in the chemical's absence. Cells are responsive to a large range of such chemical and physical triggers which can be considered 'inputs' in gene circuits where the 'outputs' are the proteins we want to make. By combining gene circuits, we can begin to create the biological equivalent of logic gates and even combinations of circuits to create, for example, oscillators (Elowitz and Leibler, 2000).

Building such circuits is not trivial, and the complexity of biological systems means that their development relies on computational modelling to help predict the design outcomes, and successful implementations are never guaranteed. The aim of synthetic biology is to simplify this process of gene circuit construction by modularizing genetic parts such that gene switches which can be swapped out and produce circuits can be combined into hierarchies of devices and systems to provide more complex functions.

Synthetic biology is based on abstraction and the hierarchical assembly of *parts*, *devices* and *systems* which "pivot around broad and abstract engineering design principles" (Mackenzie, 2010: 128). Within this definition of synthetic biology are concepts that would have been familiar to Simon. The structure of complex systems, from human brains to computers, is governed by modules which are nearly decomposable. In biology, nearly decomposable structures consist of:

> boxes within boxes . . . (e.g., a digestive system consisting of mouth, larynx, esophagus, stomach, small and large intestines, colon; or a circulatory system consisting of a heart, arteries, veins, and capillaries) as is acquired by systems that evolve by assembly of simpler systems.
>
> (Simon, 1996: 193)

We can, Simon suggests, "expect to find this separability, to a greater or lesser degree, in all large and complex systems, whether they are artificial or natural" (1996: 7).

In its aspiration, synthetic biology is also attempting to move beyond a more informal set of approaches associated with what has been described as the 'artisan craft' (what Simon might have described as 'cookbooky' methods) of genetic engineering (Calvert, 2013).

Although Simon makes little mention of molecular biology, his definition of the artificial can be applied to biology as much as any other system. In Simon's logic, a biological system can be treated as an artificial system simply by re-framing our understanding of it – not in terms of its substance but in terms of human-defined goals and describing it in terms of abstract engineering design principles.

Simon's framework does not, however, map perfectly if we delve deeper into the products of synthetic biology. In defining the 'artificial', he sets it in contrast to the 'natural', a term for which he has no definition. As noted earlier, one of the 'boundaries' between the artificial and the natural is defined in the following way: "Artificial things may imitate appearances in natural things while lacking, in one or many respects, the reality of the latter". The 'reality' of natural things is also never described. It is, at best, an intuitive category, but where intuition may suffice in distinguishing between, for example, a clock and the rolling ocean, how might we distinguish between a wild-type bacterium and a genetically modified bacterium used to produce a valuable chemical product? The genetic modification may mean that the new engineered bacteria strain lacks *in one or more respect the reality* of the wild-type bacteria, but does this mean that the engineered bacteria has lost its nature? Surely it more than 'resembles' a natural thing. By growing, subdividing and metabolizing in ways that are unrelated to its human designed functions, it retains all but a small part of its *nature*.

There is also the question of synthesis. Artificial things are synthesized by human beings. A clock is clearly a product of human synthesis, but could our modified bacteria be said to be a product of a human process? We may have provided the environment for the bacteria to grow and deleted or added a gene, but the putting together of its parts is done *in vivo* (in the living). It is largely self-synthesizing. It may become clearer using an example from Gibson et al.'s work on the chemical synthesis of a *Mycoplasma genitalium* genome (2008). In their landmark experiment, the entire genome of this simple bacterium was produced by a process of synthetic gene synthesis; that is, the genes were chemically 'printed' rather than cloned from another cell. However, thereafter, the cells divided and, aside from a few DNA 'watermarks', the measure of the project's success was the ability to transplant the synthesized genome into a cell membrane and have the cell 'reboot' and behave exactly like the natural cell by metabolizing and reproducing (Gibson *et al.*, 2010). The importance of this scientific achievement was contested, however, since the team had "not created life, only mimicked it" (Wade, 2010). The success of the final project was that the cell resembled, in almost all respects, its natural counterparts. Arguably, based on Simon's definition of the 'artificial', *M. genitalium*, since it resembles the reality of a natural thing, falls short of being artificial since it is not described in terms of human goals and functions.

The practices of synthetic biology also challenge the narratives of modularization and abstraction proposed by synthetic biologists. Synthetic biology has yet to reach its early potential; many substantial problems remain, and these prevent complex systems from being produced easily by synthetic biology methods (Kwok, 2010). There is also discomfort about the simplified models of living cells as engineered machines used by synthetic biology. Calvert, for example (2013), quotes Lazebnik (2004):

> engineering approaches are not applicable to cells because these little wonders are fundamentally different from objects studied by engineers. . . . What is so special about cells is not usually specified, but it is implied that real biologists feel the difference.
>
> (1405)

FORCIBLE CONSTRAINT V. NATURE'S OWN AGENCIES

Bensaude-Vincent and Newman suggest that there is an "absence of any clear and unambiguous terminology for distinguishing artificial and natural products in the English language" (2007: 4). However, they do offer some insights through Aristotle's faltering attempts at a definition. They describe Aristotle's definition of natural and artificial objects being a distinction between objects created with "nature's own agencies" as opposed to "forcible constraints". A tree, for example, if left to its own agency, may produce new trees. A tree can also be made into a bed, but this requires human imposition through the "forcible constraints" inherent in cutting, screwing and gluing the tree's timber. Cooking food, in contrast, uses the products of nature and the natural process of heating. Cooked food, in Aristotle's definition is, therefore, natural. Bensaude-Vincent and Newman suggest, however, that this categorization is flawed. There are many objects and materials which are made which we might consider artificial. Glass, for example, is produced using sand and "nature's own agency" of heat. Glass could be described as natural, but it feels odd to describe a glass vase as a natural object. This distinction between "nature's own agencies" and "forcible constraints" is, however, useful when we examine designing with biological systems.

In my conversations with synthetic biologists, they will often express frustration that the genetic modifications they undertake in a lab are considered different from the results of centuries of selective breeding which is itself is a form of genetic engineering, albeit relying on random mutation. Few would argue that a rose is an artificial object, but it may be the result of decades of selective breeding. A genetically modified (GM) crop, however, might be considered a patentable invention or, for an anti-GM protester, an artificial aberration of nature. The distinction is less about the object created but rather the process used to create it. In distinguishing between the practice of selective breeding and lab-based genetic modification, the former is associated with the utilization of "nature's own agencies" as opposed to the synthetic process of genetic manipulations which is considered a form of "forcible constraint".

The distinction between "nature's own agencies" and "forcible constraints" is, I would suggest, paralleled by Simon's inside/outside distinction. Returning to his example of the two clocks, the sundial and the ship's chronometer, the artificiality of both is defined by the interface between the inside and outside, where the sundial in configured in relation to the motion of the sun and the chronometer is insulated from the motion of a ship at sea. In both cases, the artificial clocks are defined by their relationship to their external 'natural' environment. It would be strange to consider improving the effectiveness of a sundial by physically moving the sun or improving the accuracy of a ship's chronometer by altering the Earth's weather patterns to minimize waves. Designers of machines are, therefore, focused on the 'inside' and its interface with an outside and un-editable nature.

This is not the case in biology. For most of human history, biological systems have been designed by shaping only the outside environment. Modern species of rose are the result of decades and longer of cross and selective breeding within environments which are carefully controlled to provide the optimum conditions for growth. Despite the fact that the result is 'genetically modified' from the original species, the manipulation has only ever been to the flower's environment – its 'outside', not its 'inside'. We don't therefore question the nature or artificiality of the rose, because we have shaped nature's own agencies. Synthetic biology through genetic modification, however, directly alters the 'inside' of the biological system and is therefore an artificial intervention. Lee describes genetic engineering of this sort as 'deep technology' which involves a 'deep' level of manipulation affecting the building blocks of matter such that the "'natural' becomes the artifactual" (1999: 255).

Preston, however, argues that this is a 'red herring'. Modern biotechnologies, she suggests, are nothing more than a footnote to a process which began with agriculture in the 'neolithic' (2013). Indeed, many of the lab practices we observe today are much more about the constraint of the outside though the creation of controlled environments. Our bacteria, whether purposefully genetically altered or not, tend to be grown in highly controlled conditions with optimum nutrition and temperatures such that the cells are able to grow and behave in predictable ways. Slight changes in these outside environments can substantially alter the 'inside' of our system. In synthetic biology, we are, to return to the ship's chronometer example, designing the clock by controlling the waves our ship is sailing through.

Machines are designed with conscious reference to an inside and outside. The computer on which I am typing is designed to minimize its interaction with the outside world – keeping out dust and water and maintaining an optimum temperature for the processor. It is, as much as possible, insulated from the 'natural world' other than through carefully controlled interfaces – the screen, the keyboard, speakers and data ports. When those interfaces are breached, as I discovered when my last computer came into contact with spilled water, the results can be catastrophic.

Biological systems work in precisely the opposite way. Their interfaces are many and varied, which is why, for most of human history, our way of designing them (and we have designed them) has been to alter their environments, using the inherent plasticity of 'nature' to shape their outcomes. In Aristotle's terms, we have used nature's own agencies.

BEYOND THE MACHINE

As we have seen, in synthetic biology, the definitions and status of natural and artificial are unclear. The design parameters which emerge from the machine metaphor, including descriptions of biological parts and hierarchies which make reference to functional goals and the distinction between inside and outside, sit well with Simon's view in *The Sciences of the Artificial*. Despite this, biological systems don't appear to conform to these frameworks; rather, we impose these frameworks on them. A key problem, suggests Mitchell (2006), is that a framework that we use to understand the world gets mistaken for how the world actually works. In other words, the map is mistaken for the territory. A problem occurs when we start mistakenly attempting to build on the map by treating natural systems in the same way as we treat synthetic systems. Catts and Zurr, for example, describe a "single engineering paradigm" in synthetic biology which is governed by precision and control and is led by an "intolerance to uncertainty" (2014). There is, in the practice of synthetic biology, a 'principle of limited sloppiness' also known as 'kludging' (finding solutions which are clumsy, dumb but good enough) (O'Malley, 2009).

The machine analogy also contains notions of efficiency and elegance which Boudry and Pigliucci suggest are misleading (2013). This analogy tends to describe biology in terms of evolutionary optimums. Living organisms are seen as being optimum machines – made efficient through evolution and natural selection. Evolution, however, is better considered a satisficer and a tinkerer. Without the opportunity of starting from scratch, evolution must make do with the materials (genetic and otherwise) at hand – "co-opting and modifying existing structures resulting in a network of interlocking and partly redundant components" (ibid, 663). There is also, they argue, within the machine metaphor a notion of a direct blueprint-like mapping between genotype and phenotype. The genome, however, provides no such 'blueprint', and cooking offers an alternative analogy whereby DNA offers instructions "but does not specify all of the details of the process, which are left to a continuous interaction between the recipe itself and the environment and ingredients that are being used" (ibid, 667).

Nicholson draws a distinction between organisms and machines, suggesting two critical differences. First, while organisms and machines operate towards certain ends; that is, both are purposeful, organisms are intrinsically responsive, as opposed to machines, which are extrinsically responsive. In other words, machines operate to ends which are external to themselves, and organisms only serve their own interests. Second, organisms, unlike machines, display only "transitional structural identity". "The constituent materials of the system change, yet the organization of the whole remains" (Nicholson, 2013: 672).

CONCLUSION

In order to facilitate a better understanding of the concept of 'designing the natural', in this chapter, I posed the question, "What happens when we make the simple exchange of terms by swapping out 'sciences' for 'designs' and 'artificial' for 'natural'?" In doing so, I set up a number of oppositions: between the natural and the artificial, between the inside and the outside of systems and between "forcible constraint" and "nature's own agencies".

The distinction made in Simon's text between the natural and artificial leads to design practices which mean framing living cells as machines. Machines are systems with a delineated artificial 'inside' which interfaces with a 'natural' outside. Machines are subject to engineering practices which seek to overcome complexity through abstraction, modularization and hierarchies of decomposable (or nearly decomposable) structures.

However, I have also shown that both in terms of outcomes and in practice, synthetic biology, which adopts the machine metaphor, also provides a challenge to it. Engineered living cells are only partially synthesized in relation to human needs and goals. In any cell which continues to grow, subdivide and metabolize, most of their nature is intact. In practice, fields such as synthetic biology, which are formed around an idealized notion of pure engineering design, end up with practices which rely more on kludging.

There is a necessity for more heterogeneous practices than are represented by the engineering ideal, because the inside/outside distinction is no longer tenable in biology. Modern synthetic biology is defined by our desire to tinker with the 'inside of cells' to alter their nature – mainly through genetic manipulation. However, millennia of 'biotechnology' in agriculture and practices such as selective breeding have shaped species through the control of the 'outside' environment. Even in modern synthetic biology, with the tools available to alter the inside of cells, much of the work is still about the control of the environment rather than the cell itself. Tinkering with nature through, for example, genetic manipulation in the artificial environment of the lab, is considered forceful in contrast to selective breeding, which uses nature's own capacities. The outcomes of the latter, the rose, for example, are considered natural in contrast to the artificial intervention made through genetic engineering of the 'inside' of the cell, even though the genetic 'distance' between a modern rose and its ancestors may be more significant than a wild-type species which has undergone a single gene edit. The latter has changed the *inside* of the rose cell and its interface with the natural world, whereas the former has shaped the rose through the cultivation of the *outside* environment.

The machinic metaphor, then, is limited, as are distinctions between the artificial and the natural. These metaphors are, to some extent, inevitable; as Lakoff and Johnson point out: "[m]etaphorical thought is what makes abstract scientific theorizing possible" (1999: 128). In my own experience, there is a utility in learning biological principles through references to technologies which I already understand. And, as I discussed at the beginning of this chapter, I have found it useful to invoke imperfect analogies when teaching biology to design students. It is easy to criticize but less easy to replace. Designing the natural still requires us to develop methods and to anticipate outcomes which are unpredictable.

The answer to the challenge of Bio Design methods lies somewhere between outside and inside and between the natural and the artificial, between nature's own agencies and forcible constraints. The machine metaphor may be wrong and set up design frameworks which are not applicable, but it is difficult to replace them with something else. I will use the rest of this book to investigate possible ways forward in the context of assembly and fabrication, but looking at how biological systems organize matter and how we as designers can intervene.

REFERENCES

Bensaude-Vincent, B. and Newman, W. R. (2007) 'Introduction: The Artificial and the Natural: State of the Problem', in Bensaude-Vincent, B. and Newman, W. R. (eds) *The Artificial and the Natural: An Evolving Polarity*. Cambridge, MA: MIT Press, pp. 1–19.

Boudry, M. and Pigliucci, M. (2013) 'The Mismeasure of Machine: Synthetic Biology and the Trouble with Engineering Metaphors', *Studies in History and Philosophy of Science Part C: Studies in History and Philosophy of Biological and Biomedical Sciences*. Elsevier Ltd, 44(4), pp. 660–668. https://doi.org/10.1016/j.shpsc.2013.05.013.

Calvert, J. (2013) 'Engineering Biology and Society: Reflections on Synthetic Biology', *Science Technology & Society*, 18(3), pp. 405–420. https://doi.org/10.1177/0971721813498501.

Campos, L. (2009) 'That Was the Synthetic Biology That Was', in Schmidt, M. *et al.* (eds) *Synthetic Biology*. Dordrecht: Springer Science, pp. 5–21. https://doi.org/10.1007/978-1-62703-625-2.

Catts, O. and Zurr, I. (2014) 'Countering the Engineering Mindset: The Conflict of Art and Synthetic Biology', in Ginsberg, A. *et al.* (eds) *Synthetic Aesthetics: Investigating Synthetic Biology's Designs on Nature*. Cambridge, MA: MIT Press, pp. 27–38.

Dade-Robertson, M. (2011) *The Architecture of Information*. London: Routledge.

Elowitz, M. B. and Leibler, S. (2000) 'A Synthetic Oscillatory Network of Transcriptional Regulators', *Nature*, 403(6767), pp. 335–338. https://doi.org/10.1038/35002125.

Endy, D. (2005) 'Foundations for Engineering Biology', *Nature*, 438(7067), pp. 449–453. https://doi.org/10.1038/nature04342.

Gibson, D. G. *et al.* (2008) 'Complete Chemical Synthesis, Assembly, and Cloning of a Mycoplasma Genitalium Genome', *Science (New York, N.Y.)*, 319(5867), pp. 1215–1220. https://doi.org/10.1126/science.1151721.

Gibson, D. G. *et al.* (2010) 'Creation of a Bacterial Cell Controlled by a Chemically Synthesized Genome', *Science (New York, N.Y.)*, 329(5987), pp. 52–56. https://doi.org/10.1126/science.1190719.

Hillis, D. (2016) 'The Enlightenment is Dead, Long Live the Entanglement', *Journal of Design and Science*, 1(1).

Huppatz, D. J. (2015) 'Revisiting Herbert Simon's "Science of Design"', *Design Issues*, 31(2), pp. 29–40. https://doi.org/10.1162/DESI_a_00320.

Ito, J. (2016) 'Design and Science: Can Design Advance Science, and Can Science Advance Design?', *Journal of Design and Science*, 1(1).

Kay, L. (1991) *Life as Technology: Representing, Intervening and Molecularizing, Working Papers, Programme in Science Technology and Society*. Cambridge, MA. Available at: http://books.google.com/books?hl=en&lr=&id=605eE1hMW3kC&oi=fnd&pg=PA87&dq=Life+as+technology-+representing,+intervening+and+molecularizing&ots=VS_bIuc2nd&sig=zMfNh-UbqSnt3gwnrHwgp-XlKpE.

Kingsland, S. E. (1991) 'The Battling Botanist: Daniel Trembly MacDougal, Mutation Theory, and the Rise of Experimental Evolutionary Biology in America, 1900–1912', *Isis*, 82(3), pp. 479–509. https://doi.org/10.2307/233227.

Kitney, R. (2012) *Synthetic Biology Scope, Applications and Implications*. London: Royal Academy of Engineering. https://doi.org/10.1021/sb300001c.

Kwok, R. (2010) 'Five Hard Truths for Synthetic Biology', *Nature*, 463(7279), pp. 288–290. https://doi.org/10.1038/463288a.

Lakoff, G. and Johnson, M. (1999) *Philosophy in the Flesh*. New York: Basic Books.

Lazebnik, Y. (2004) 'Can a Biologist Fix a Radio? – or What I Learned While Studying Apoptosis', *Biochemistry*, 69(12), pp. 1403–1408. https://doi.org/0006-2979/04/6912-1403.

Lee, K. (1999) *The Natural and the Artefactual: The Implications of Deep Science and Deep Technology for Environmental Philosophy*. Oxford: Lexington Books.

Mackenzie, A. (2010) 'Design in Synthetic Biology', *Biosocieties*, 5(2), pp. 180–198. https://doi.org/10.1057/biosoc.2010.4.

Mitchell, S. D. (2006) 'Modularity – More Than a Buzzword?', *Biological Theory*, 1(1), pp. 98–101. https://doi.org/10.1162/biot.2006.1.1.98.

Nicholson, D. J. (2013) 'Organisms = Machines', *Studies in History and Philosophy of Biological Science and Biomedical Sciences*. Elsevier Ltd, 44(4), pp. 669–678. https://doi.org/10.1016/j.shpsc.2013.05.014.

O'Malley, M. (2009) 'Making Knowledge in Synthetic Biology: Design Meets Kludge', *Biological Theory*, 4(4), pp. 378–389. https://doi.org/10.1162/BIOT_a_00006.

Pauly, P. J. (1987) *Controlling Life: Jacques Loeb and the Engineering Ideal in Biology*. Oxford: Oxford University Press.

Preston, B. (2013) 'Synthetic Biology as Red Herring', *Studies the History and Philosophy of Biological and Biomedical Sciences*, 44(4), pp. 649–659. https://doi.org/10.1016/j.shpsc.2013.05.012.

Protzen, J.-P. and Harris, D. J. (2010) *The Universe of Design: Horst Rittel's Theories of Design and Planning*. London: Routledge.

Schon, D. A. (1984) *The Reflective Practitioner*. London: Basic Books.

Simon, H. A. (1996) *The Sciences of the Artificial*. Cambridge, MA: MIT Press.

Wade, N. (2010) 'Researchers Say They Created a "Synthetic Cell"', *The New York Times*. Available at: www.nytimes.com/2010/05/21/science/21cell.html?_r=0 (Accessed: 4 December 2017).

Watson, J. D. and Crick, F. H. C. (1953) 'Molecular Structure of Nucleic Acids', *Nature*, 171, pp. 737–738.

The logic of living assembly

Chapter 3

BIOLOGICAL CONSTRUCTION: BETWEEN ASSEMBLY AND FABRICATION

Having developed a broad overview of the relationship between the *natural* and *artificial* in Chapter 2, Chapters 3 and 4 will focus on the question of how biological systems construct material structures and how we can guide this process. In doing this, I want to move away from the machine metaphor and the idea of an inside and outside of biological systems.

In this first chapter, I am going to stay away from specific examples of material construction (although some will be alluded to), as this is not, as I indicated in the Introduction, a textbook on biology or biomaterials. My aim is to consider biological systems the basis for material construction.

In construction, it is helpful to distinguish between *assembly* and *fabrication*. A distinction between these processes in traditional manufacturing can be illustrated with reference to a piece of flatpack furniture. Fabrication is the process by which the parts are made. Imagine, for example, a flatpack bookcase made from medium-density fibreboard. The panels which make up the bookcase shelves and structure are fabricated from reconstituted wood, which is then cut into parts of a predetermined length. These parts have holes drilled into precise locations to allow for the screws and bolts. These manufactured parts are simple, in terms of their form and material composition, compared to the final object. The shelves are then *assembled* from these simple parts by following instructions which designate their position and order of assembly. In the case of flat-packed furniture, the process of assembly and the manufactured parts are, in theory, relatively simple to allow for assembly to be conducted by a non-expert, for the process to be completed in a small number of steps and for the parts to be fabricated cheaply. In more complex assemblies such as the manufacture of a car or a building, there will be a hierarchy of assembly with simple manufactured parts assembled in stages.

In the cases of both simple and complex manufactured products, we would tell the story of manufacture in terms of the fabrication of parts and then their assembly. However, biological systems operate according to a different logic. If we observe the production of materials from any biological system, it is impossible to make a clear distinction between fabrication and assembly. In fact, I will argue here that, in understanding living cells as a medium for making material structures, we need to start and end with assembly. All biological matter is assembled from simple parts to form more complex parts, which, in turn, form more complex parts and so on. We could say that proteins are fabricated from simpler molecules, amino acids, and then assembled into protein complexes which can form some structural biomaterials. However, this delineation of fabrication and assembly is arbitrary. We could just as easily describe cells as fabricated parts which are subsequently assembled into multicellular organisms. While biological systems exhibit a functional and structural hierarchy, there is a continuum of assembly and not a series of clearly delineated and separate stages.

In this chapter, rather than defining the *inside* or *outside* of a biological system, as we do when describing artificial systems, I will describe biological assembly, across scales and complexity, in terms of the characteristics and interaction of four parameters: *matter*, *energy*, *force* and *space*. Importantly, these parameters contain, through their interactions with one another, *information* such that matter is patterned into increasingly complex configurations. We need to establish, therefore, how the parts are instructed to fit together in different ways and how we, as designers, can alter the information content to alter the process of assembly.

We can, for example, describe the process of putting together a flatpack bookcase using this approach. *Matter* refers to the substance of the bookcase, in this case chipboard sheets and steel bolts and screws. *Energy* is required to move the parts and is translated into directed *forces* which push the parts together. Information must exist, in this case, to pattern the parts (matter) and forces (energy) to assemble the parts in the right configuration. In the case of a flatpack bookcase, *information* is encoded through the form of the parts themselves – through their size and shape and the locations of holes for bolts and screws and grooves which locate connecting parts. *Information* is also contained in the instruction manual, which, in turn, through a human interpreter, communicates a sequence of actions – directing the forces applied to move the parts to be assembled in specific ways. In other words, the instruction book defines a sequence of pathways to direct human energy to exert forces to move and locate the parts in the correct pattern and in the correct sequence.

While this is a convoluted way of describing the assembly of a piece of flatpack furniture, it does allow us to ask an important question about its design. Imagine an alien species that, in viewing the earth, doesn't make a distinction between human beings as agents and flatpack bookcases. What they observe from their spacecraft is an ongoing process of *self-assembly*. Interactions of different elements of matter produce bookcases spontaneously. For some reason, flatpack bookcases are very useful for our alien species, but they don't possess the necessary knowledge to understand how to produce them (hex keys have not been invented on their planet). Instead they want to intervene to allow this spontaneous process to happen but to alter the outcomes such that the resulting furniture matches their specifications. Depending on the tools at their disposal, they would seek ways of intervening in the process by changing the information in the assembly system. They could, for example, make changes to the assembly instruction sheets, adding in steps or altering the order of assembly. They could, if they had the means to do so, alter the fabrication of the parts such that holes and slots are cut in different places. They may do a combination of those things, but the assembly of the bookcase would still happen autonomously – outside their direct control. They would simply harvest the bookcases once they were made.

Davies (2016) describes biology as an "alien technology". There is a lot we don't know about how living cells work. We can, as I described in Chapter 2, synthesize the genome of a simple organism entirely synthetically, insert it into an empty cell membrane and have the cell 'boot-up' as demonstrated by Craig Venter's team with the bacteria *M. genitalium* (Gibson *et al.*, 2008), but it doesn't mean we know how all the genetic parts work. In this case, we know where the information source is which will enable a viable cell to be produced. Like our aliens, we don't have the tools to construct the bookshelves, but we might know how to manipulate the *information* which enables the shelves to be produced.

LOGICS OF ASSEMBLY IN ALIEN TECHNOLOGY

When we ask the question as to where the information is to assemble traditional design products, we tend to point to external representations – drawings, models, schedules of work and so on. In architecture, for example, we don't see the architect's role as making buildings but rather in "making mediating artifacts which make significant building possible" (Perez-Gomez and Pelletier, 2000: 7). The information required to construct buildings resides in a rich array of models and drawings which inform construction, not in the elements of construction themselves. The example of a piece of flatpack furniture is an interesting one because the designers must reduce the complexity of the instruction set compared to, for example, the blueprints and tacit knowledge which may be used as instructions in a factory with assembly undertaken by expert furniture assemblers. The flatpack designers also have no control over the space in which the furniture will be assembled, unlike a factory setting, where the environment is purpose built for the task. Through the machining of specific parts, information contained in the parts themselves is increased so that the parts will fit together in a limited number of ways to reduce, although from personal experience not eliminate, the possibility for assembly error as well as simplifying the instructions.

If we turn back to biology, however, the question of where the information is located becomes much more complicated to answer. There is no 'external' representation such as a blueprint for a biological system. There are known information carrier molecules such as DNA, but these, in themselves, don't provide the assembly instructions for a cell, let alone a multicellular organism. In her lecture 'The Physics of Life' delivered to the Royal Institution, Sylvia McLain describes DNA as "like the middle management of molecules" (2017). Information in biological systems is distributed. It does not reside in a blueprint or in a single process. There are, however, logics of assembly, and information resides in multiple places. By observing how biological assembly works at multiple scales and how information shifts from one parameter to another, we can depict the process of biological assembly as a continuum starting with simple molecules and the self-assembly of atoms that we see in chemistry and growing in both scale and complexity to the assembly of organic molecules and complex structures within a cell and to the assembly of cells themselves.

It is impossible to be exhaustive in discussion of these processes, but I propose to use generic examples (which will be related to chemical and biological processes) and schematic representations, focusing on the information content of each example. How do the parts know how to assemble, or how are they instructed to form into more complex structures? In other words, where is the information?

CHEMISTRY: BOUNDING, CONCENTRATION AND THE PATTERNING OF FORCES

Chemical processes involve the assembly of atoms into molecules through molecular bonding. Described as a process of assembly where simple 'parts' interact, we can build a picture of the necessary conditions for assembly to occur.

Figure 3.1a depicts a space in which elements of matter – which could represent atoms but which I will describe as parts – are depicted as spheres. In Figure 3.1b, the system is energized, meaning the elements are now animated. In the diagram, the elements of matter are performing a random walk as the atoms or molecules jostle with each other. In this case, no assembly will occur. In fact, this system will tend to lead to disorder, since the parts of matter will tend to move apart. In Figure 3.1c, the space is now bounded, meaning that the parts will remain in the space defined by the boundary. The elements of matter will not assemble, but the boundary introduces information into the system. While the boundary represents a simple physical barrier, it is also an *instruction* for the parts not to move out of a defined space. While not assembled, the elements of matter will tend, on average, not to move further apart. In Figure 3.1d, we introduce force attractors onto the elements of matter. These are described in the diagram as magnets but are, in fact, molecular bonds. These attractors are arranged on the surface of the elements of matter at particular angles. Information is constituted by the patterning of these elements across the surface of the particles and the angle of intersection. This pattern informs the possible assemblies so that, as each part moves randomly through space, it will eventually connect with other parts and form a stable structure (unless other forces are applied), as seen in Figure 3.1e. We see this process of structuring in the arrangements of atoms into molecules through molecular bonds.

In this schematic illustration of molecular assembly, the information in the system is held in the bounding of the space, meaning that the parts cannot drift apart, and in the patterning of bonds. Energy in the system is applied as a random force through the system so the parts only connect when they collide. We can, however, increase the chances of their assembly by either increasing the energy in the system, making the parts move through space faster, or increasing the concentration of parts by decreasing the size of the bounded space or increasing the number of parts within

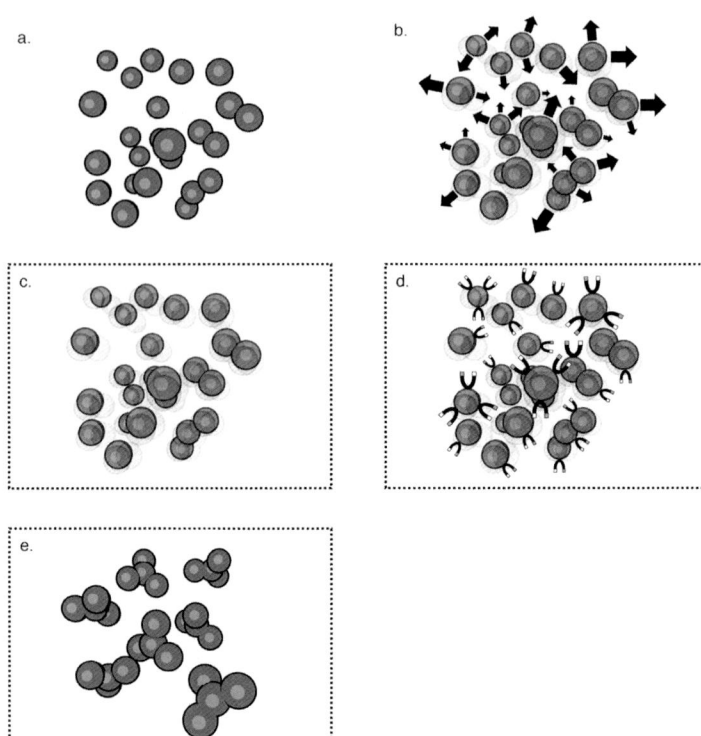

Figure 3.1 The minimum conditions necessary for molecular assembly include elements of matter (a) which are energized to move randomly through space (b) and constrained to a bounded space (c) and include patterned covalent bonds [represented in (d) by horseshoe magnets] which cause the matter to be assembled into groups of atoms (e).

the defined space. Altering the amount of energy in a system (for example, by heating) or the concentrations of parts are common tools in chemistry for increasing the speed of reactions.

INSIDE A CELL: PATTERNING OF PARTS AND THE CONSTRAINT OF FORCES

While 'simple' chemistry occurs within cells, a defining feature of intracellular (within the cell) chemistry is the formation of complex organic molecules which would not assemble (or at least would be very unlikely to assemble) spontaneously without a mediating process. It follows that to build more complex molecular structures, we need to increase the amount of information in our system.

In Figure 3.2, we see a chain of parts which are held together and able to articulate at hinges. The chains are able to move freely within the bounded space, and their movements are random. However, parts of the chain are only able to move along specific paths in relation to one another (Figure 3.2a). The parts also contain force attractors which will tend to repel or attract the force attractors on other parts of the chain. The chain will tend to fold, therefore, into predictable patterns and remain stable once it has folded (Figure 3.2b).

Information in this system is encoded in the sequence of parts, the patterning of forces across their surface and the constraint placed on the movement of parts by the hinge.

In Figure 3.3, we see aggregations of parts making 3D shapes. These parts are still held within a finite space and are subject to

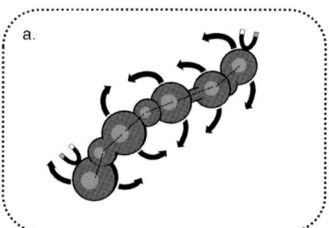 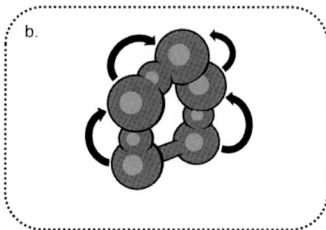

Figure 3.2 A chain of parts (a) in which random motions are constrained and forces are patterned – in this case, at the ends of the chain, such that the chain will tend to fold into specific geometries (b).

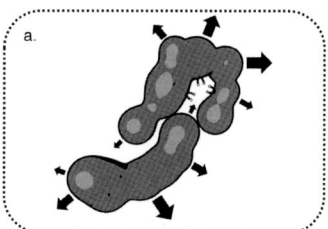 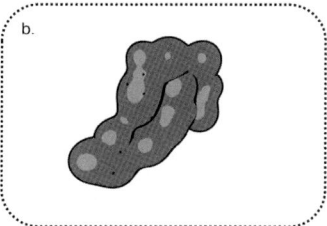

Figure 3.3 The assembly of jigsaw-like parts where a combination of conformational geometry and the patterning of forces (a) causes randomly moving parts to assemble when they meet in the right configuration (b).

energy which causes the parts to move randomly. Their surface is patterned by attractors, and aggregations have different shapes which lead to complementary geometries, like jigsaw pieces, such that, if they meet in the right conformation, they will tend to stick together to make larger, more complex agglomerations (Figure 3.3b). While random collisions may occur between the *jigsaw pieces*, the strongest and most stable patterns only occur when pieces connect with maximum conformation. This means that such structures can form using comparatively weak attractors. Information here is held in the geometry of the agglomerations and the patterning of forces across the active surfaces of the jigsaw puzzle.

The assembly methods described in Figures 3.2 and 3.3 can be found, for example, in protein assembly. The folding of the proteins often leads to complex structures (see, for example, Figure 3.4) and, while the process is understood, predicting the formation of a protein molecule from a sequence of amino acids is very difficult, and this is known as *the protein folding problem* (Dill and MacCallum, 2012). Proteins, however, all follow a common assembly pathway starting with primary assembly, consisting of a chain of amino acids which fold into a secondary structure, either a helix or sheet, and then further fold into a tertiary structure and sometimes a quaternary structure where more than one protein molecule will assemble together. It is also worth noting that, unlike my highly simplified example, proteins don't have directional hinges as such but fold into specific conformations by orienting amino acids in water with hydrophobic (repelled by water) molecules pushed to the centre of a protein assembly and hydrophilic (water-attracting) molecules forced to the outside.

Using our knowledge of an amino acid sequence, it should be possible to design novel proteins by programming specific chains of

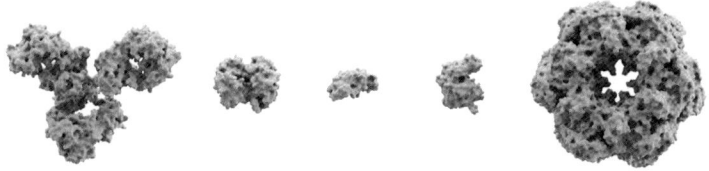

Figure 3.4 The molecular surface of three common proteins and protein complexes. From left to right, immunoglobulin, haemoglobin, insulin, adenylate kinase and glutamine synthase (www.wikiwand.com/en/Protein).

amino acids, and there is a body of research into this area (Gibney, Rabanal and Dutton, 1997).

The process illustrated in Figure 3.3 is found in protein assemblies where a number of independent protein molecules will fit together to form protein complexes. These parts might include two different types of protein: 'heterodimers', which are fitted together asymmetrically, and 'homomers', which are made of two or more identical proteins, often symmetrically arranged. These complexes form through 'avidity', which is a weak patterning of forces where multiple points of contact in a protein assembly make up for the fact that the individual forces are weak. The weak forces involved also act as a self-checking mechanism, as the molecules will tend, in energetic conditions, to break apart because they are only stable when they are assembled correctly. Through this assembly method, it is possible to form macro structures, with different assemblies forming linear, cubic and lattice patterns (Norn and André, 2016).

In Figure 3.5, I show a variation on the assembly method illustrated in Figure 3.3. However, in this case, we have three interacting parts where one part acts as a template for the other two. It, in turn, is

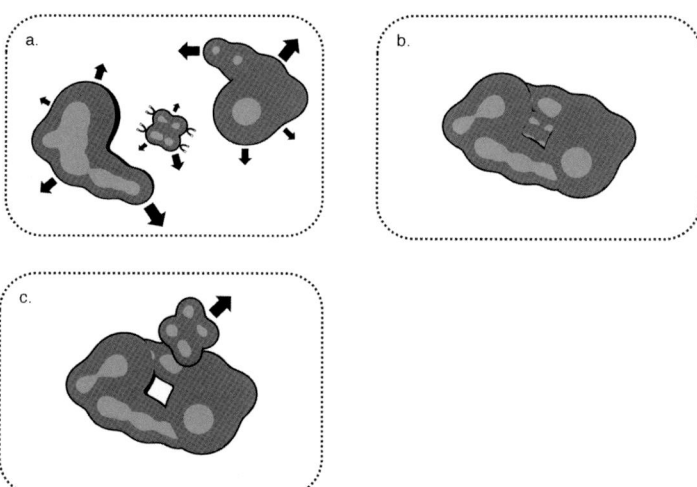

Figure 3.5 Templated assembly where one part acts as a template for two others but does not form part of the final assembly.

detached, leaving the two remaining parts to form the assembly. In this case, information is not only held in the parts to be assembled but in an external agent which itself does not form part of the final assembly.

These independent information carriers are essential in cell biology. Proteins themselves often act as the templates for metabolic interactions by shepherding simple molecules to catalyze otherwise unlikely reactions. Some protein complexes such as insulin form with the aid of a template protein, while others are often folded with the help of chaperone proteins. The idea of an information carrier molecule acting as a template is essential to many biological assembly processes and is most notably found in the 'central dogma', which consists of a process of *transcription*, where DNA provides a template for the assembly of messenger ribonucleic acid (mRNA), and *translation*, where mRNA is used to assemble a sequence of amino acids using a complex molecule known as a ribosome.

This capacity of cells to utilize templates enables them to store information separately from assembly, and it is a key way in which we engineer biological systems. Fields such as synthetic biology rely on our capacity not to alter the assembly of proteins directly but to alter DNA sequences which will act as the template for mRNA, which, in turn, is a template for amino acid assembly. DNA gives us access to part of the information required to alter biological assembly.

The examples described in Figures 3.1, 3.2, 3.3 and 3.5 are highly simplified versions of assembly processes which take place inside a cell. They are also not exhaustive, but by looking for the information content in these examples of assembly, we can make some general observations.

For the assembly of non-organic molecules, information for assembly is predominantly held in the parts themselves through the patterning of forces and their geometry. The motions of the elements are, in the main, chaotic, so there is no patterning of the energy in the environment (no invisible hand) to direct the elements in a particular way. Assembly, therefore, depends on the chance encounter of two or more parts moving randomly in space, colliding and attaching to one another. The instruction for assembly is held in the geometry of the combining elements and the patterning of forces which will hold complementary elements together.

Assembly using this process of random encounter is possible at molecular scales within the cell because the elements, small molecules and larger macromolecules, are very light, and their environment is, compared to their size, highly energetic, so they move round the cell at rapid speeds. There are also a large number of them. To give some sense of the scale of these interactions, there are, within an *Escherichia coli* bacterial cell, approximately 2.36×10^6 protein molecules at any one time moving around the cell (Kühner *et al.*, 2009) and they take only fractions of a second to traverse the whole cell (Philips *et al.*, 2012). Put simply, there is so much 'stuff' going on in the cell, the sheer numbers and the speed with which things are moving make improbable interactions between parts into certainties. Throw a dice once and you have a 1:6 chance of landing a 6. Throw it 6000 times and you can bet with virtual certainty that you will throw a 6 at least once, and you can predict with some confidence that you will throw a 6 close to 1000 times. There is a similar principle in the cell. The chances of two complementary components crashing together is small, but the parts are colliding with such regularity because of their speed that, given significant numbers of each part within the cell (concentration), these collisions will inevitably form useful assemblies some of the time. In some cases, the parts are also guided by signal peptides which direct the movement of proteins.

As complexity grows, template elements become increasingly important. The information in templates is held outside the parts (for example, amino acids) to be assembled. Templates also decrease the number of necessary chance encounters needed for complex structures (such as proteins) to be produced.

The probability of these interactions is also increased if some of the components are shared across different processes, as evidenced, for example, by the limited amino acid set which constructs proteins. This method of assembly, where the information is predominantly contained within the elements to be assembled, is, I would contend, specific to the scale of molecules and macromolecules bounded by a cell. If we attempt to repeat the assembly of elements through their arbitrary interaction at the macro scale, the energy required and the time it would take to assemble are substantial. Compare, for example, the self-assembled model chair described by Papadopoulou, Laucks and Tibbits (2017) which consisted of six components agitated within a water-filled tank. Magnets at the joints of the assembly were patterned such that the chair could only be assembled in pre-specified configurations, and the components were agitated so that the pieces moved randomly in the water. The pieces eventually came together after seven hours. If this assembly were to be made more quickly, either more energy would have to be put into a system or more parts or both, assuming that the parts could withstand the higher amounts of energy involved in the collisions without breaking. Specific data on the size of the tank and the relative size of the chair parts is not available, but, from the published material, the chair parts would appear to be approximately 1/8th of the length of the tank. Bacterial cells can vary significantly, but an *E. coli* cell, for example, might measure 2 micrometers across. A protein molecule which will form part of a larger assembly could be in the region of 0.01 micrometers. In other words, the cell is 200 times longer than an individual protein molecule. Repeating the fluid assembly chair experiment with chair parts of 0.5 meters in length would, if they were placed in a vessel sized equivalently to a cell, require a tank 100 meters long. Consider the amount of energy required and the time it would take for the pieces to self-assemble in this context.

This scaling problem must be overcome even in the cell. As complexity increases, we see an increase in information to decrease chance encounters, make the assembly of parts more predictable and increase the combinatorial possibility of a relatively few simpler parts. In such complex arrangements, information is found in the pattern of constraints placed on the forces such that parts move in predictable directions, as in the folding of the chain of amino acid molecules, or where information content resides in templates which do not form part of the final assembly.

MULTICELLULAR ASSEMBLY: PATTERNING OF ENVIRONMENT

As we move up through length scales, we can begin to treat cells as parts in their own right, assembling to become multicellular organisms or, in the case of prokaryote cells, communities of collaborating cells (Shapiro, 1998). The sheer variety of multicellular biological patterning is a source of scientific wonder and artistic inspiration – sometimes both together (see, for example, von Goethe,

2009; Haeckel, 2012). Given this variety, the scale of the problem in describing multicellular assembly seems huge. However, as with assembly within the cell, the processes of multicellular assembly can be simplified into rules which depend on the presence of information in the parts (cells) themselves and the energy and forces present in their environment.

A useful starting point for this exploration is to note that we don't, in the main, see the cells energized and moving randomly through space waiting for collisions to happen.

In the study of multicellular organisms, the process by which cells assemble into patterns is known as morphogenesis. Davies provides some useful summaries of the processes involved in morphogenesis (Davies, 2005, 2008) and identifies ten mechanisms described as apoptosis (selective cell death), proliferation, cell fusion, locomotion, chemotaxis (whereby cells follow chemical gradients), adhesion, sorting, epithelial-mesenchymal transition (EMT) (the process by which cells which are stuck together to form epithelial tissues become detached and able to migrate), folding and mesenchymal-epithelial transition (MET) (where mobile cells stick together to form epithelial tissues). In multicellular organisms, tissues are formed through some combination of these mechanisms which are choreographed in time and space.

We can further abstract from Davies' mechanisms of morphogenesis by posing the same question used in the last two sections. Where is the information for assembly? Here we ignore the complexity of the intracellular (inside the cell) interactions and see the cell as the part. Again, I don't offer an exhaustive review but rather an illustration of the difference, in this case, between assembly between cells and assembly processes within cells. As with the example of molecular assembly, intercellular assembly requires the interaction of *energy*, *matter*, *force* and *space*.

Some assembly information is held within the cell itself, and there are, essentially, six possible types of information held by a cell:

- the cell's shape,
- the cell's orientation,
- the cell's stickiness (its likeliness to adhere to the surface of other cells or materials),

- edge detection,
- force detection,
- signalling chemicals that can be released outside the cell and used to communicate with other cells.

By modifying these sources of information, cells assemble by patterning their environment over time, where we define the

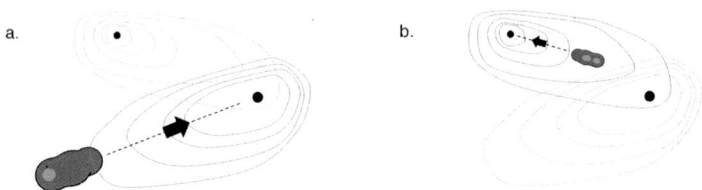

Figure 3.6 In the process of chemotaxis, a cell navigates following a chemical gradient towards a source (indicated by the black circle). In morphogenesis, cells can be directed along complex paths by a series of such chemical gradients.

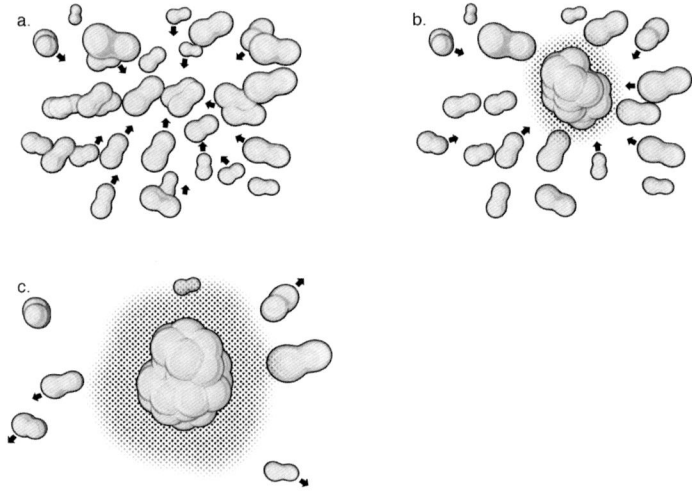

Figure 3.7 Cells forming clusters (a). As the cluster becomes denser, a chemical signal is released (b), and, as the cells reach a critical mass, the signalling molecules prevent further clustering (c).

environment as the space outside the cell and other cells. For example, some types of tissue form as a result of free cells following chemical gradients (chemotaxis), navigating along specific paths towards destinations which are indicated by higher concentrations of a signalling chemical (Figure 3.6).

Using chemical signalling, cells can also condense – sticking to one another until they reach a critical mass using a chemical signalling process called quorum sensing so that, as the cells aggregate, the buildup of an extracellular chemicals will eventually stop the surrounding cells from joining the cluster **(Figure 3.7)**.

Cells which have already formed themselves into a sheet (epithelial cells) can, during their development, close holes by detecting the leading edge (i.e. cells which have an edge which is not in contact with other cells) and moving across a surface to close gaps (Figure 3.8).

Cells formed into sheets can also be subject to mechanical forces. For example, as a tissue develops, tension can form (Figure 3.9a), causing the cells either to stretch (Figure 3.9b) or to reorganize by slipping past one another to form a new pattern (Figure 3.9c).

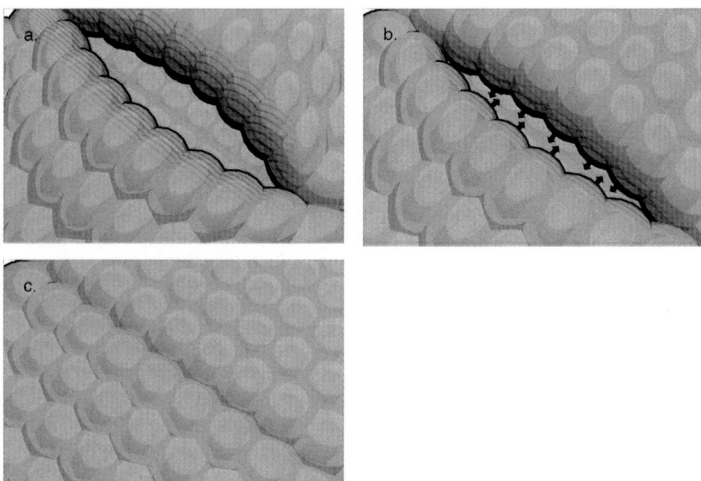

Figure 3.8 Cells formed as a series of layered sheets can detect breaks in their surface. The leading edge of the cells on each side of the break can migrate across a surface to pull the seams of the break together.

Cells can also respond by moulding by their environment through chemical signals. By softening extracellular material, cells can grow into the softened area created by chemicals produced locally by the cells (Figure 3.10). Davies (2005) gives the example of a hair follicle which forms as a process triggered by chemical signals causing an extracellular gel to soften, allowing cells to fill the area of lower viscosity underneath the forming follicle.

These are limited examples, but they all illustrate the approach taken in multicellular assembly, which can be summarized as patterning of the environment. Patterns which form between cells in multicellularity occur because cells communicate with one another through chemical signals and mechanical stresses. In design terms, the patterning of the environment has important implications, and much research into tissue engineering and regenerative medicine, for example, is concerned with the patterning of cells by developing surfaces and scaffolds to encourage cells to form structures (Brown, 2013).

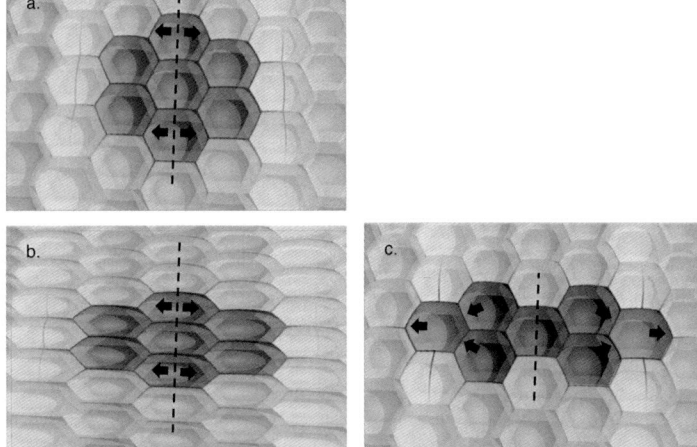

Figure 3.9 As tension develops across a sheet of cells (a), the cells can respond by changing their shape by stretching (b) or reorganizing themselves into a different configuration (c).

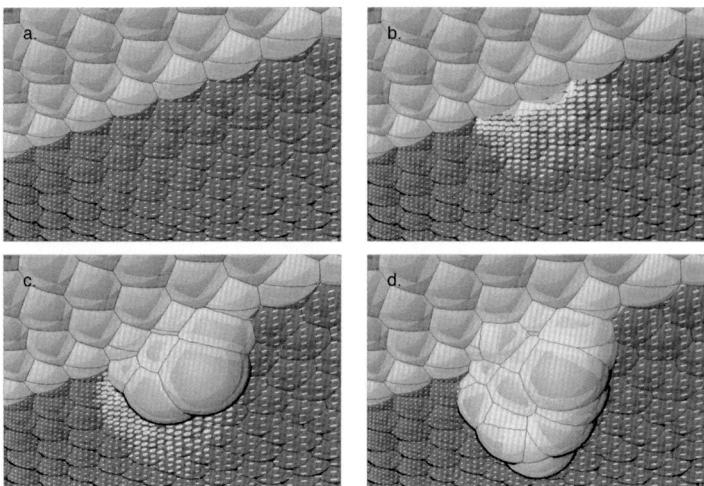

Figure 3.10 Cells can form into 'moulds', created when a dense substrate is broken down by a local chemical diffusion and cells grow to fill in the space.

EMERGENCE: INFORMATION FOR FREE

When many parts are interacting together with simple rules of assembly, we often see patterns which are neither encoded in the parts themselves nor in their environment (Holland, 2014). Instead, these systems exhibit emergent behaviour. In my earlier descriptions of the logic of assembly, there is little explanation of how, for example, a bacterial colony can grow into complex swirls and fractal patterns (Ben-Jacob, 1997), let alone how complex tissues form in multicellular organisms.

If the design of biological assembly depends on us identifying where information is located, then emergence provides a problem because, as I will show, in emergent systems, some information is not present in the system of assembly until the parts have assembled. We can create synthetic systems which exhibit emergent behaviours, including, for example, computer simulations such as Turing patterns (Turing, 1952), and these models are often used as a way of describing the formation of patterns in nature.

We can understand more about emergence by using a theoretical example. Figure 3.11 uses a simplified version of a system known

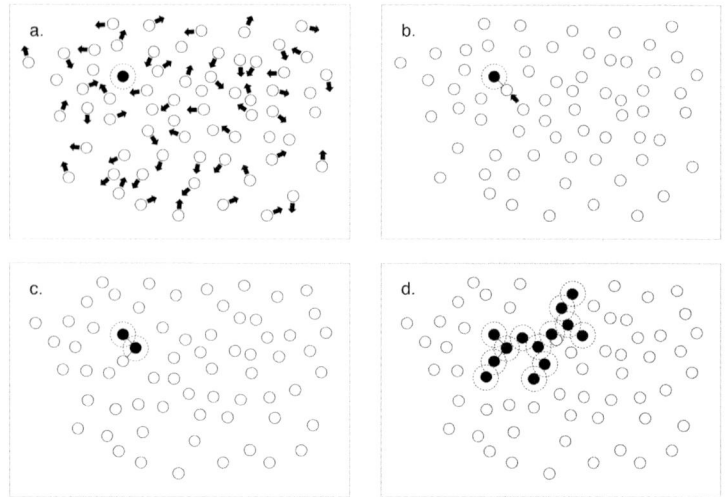

Figure 3.11 The rules for a formation of a DLA pattern where one cell represents a static 'seed' (a) and the others are in random motion until they come into proximity to the seed (b) and stop moving and become seeds themselves. (c) Through this interaction, the cells eventually form branching patterns (d).

as diffusion limited aggregation (DLA). In this system, parts are energized but not directed, moving randomly in a bounded space. One of the parts is designated as a seed (shown as the black circle) so that, if another particle comes close to it (its area of influence is shown here as the dotted circle), it will stick to the attractor particle. Once it is 'stuck', it, in turn, also becomes an attractor.

If we simulate this behaviour with enough 'energy' that the parts are agitated, we don't simply see a random aggregation of parts. Instead, distinctive branching patterns are produced. If we run the simulation more than once, given the same instructions, different (although similar) patterns will form (Figure 3.12), and we can produce computer code which generates these patterns with a few simple rules. Information needs to be encoded in the parts which move randomly in a bounded space, and they need to be aware of their proximity in relation to the seed(s) and to know whether they themselves are seeds. Once a particular pattern has been generated, however, if we try to develop a system in which the exact same

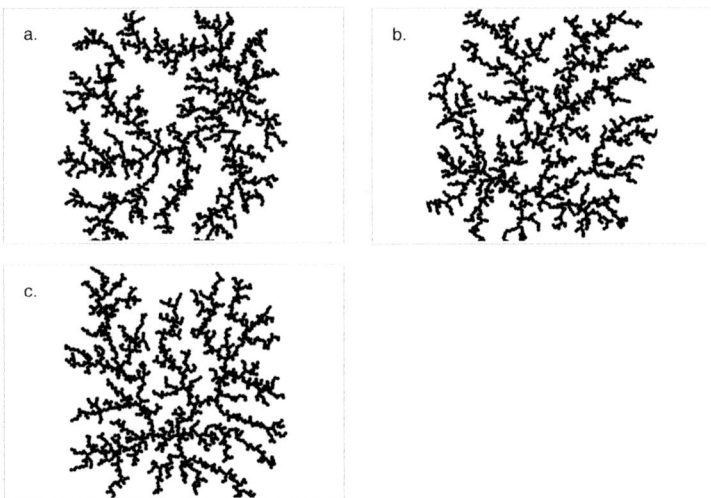

Figure 3.12 Outputs from a computer simulation of a simple diffusion limited aggregation system, each started with the same parameters but with random starting positions and movements of the cells in two dimensions.

pattern is replicated, then we need to encode substantial additional information on the relative location of each part. The parts are no longer generic, but information would need to be added to allow the parts to assemble in specific ways with precise relationships to each other, or, alternatively, the parts would need to be made to move in specific directions to create an assembly sequence which matches the one produced in the random simulation.

Emergent systems are, by their nature, more difficult to design with because we cannot ask the simple question 'Where is the information?' since the patterning of parts gains complexity (and hence information) as the system develops. This is not to say that we cannot influence the outcome. In the DLA example, the pattern of assembled parts always turns out differently because the parts move randomly through space and so their collisions are also random. However, the patterns which emerge all have a similar character, and we can influence the patterning process by altering the behaviour of the parts or the environment. For example, increasing the density of particles will alter the character of the pattern produced (Figure 3.13).

The logic of living assembly

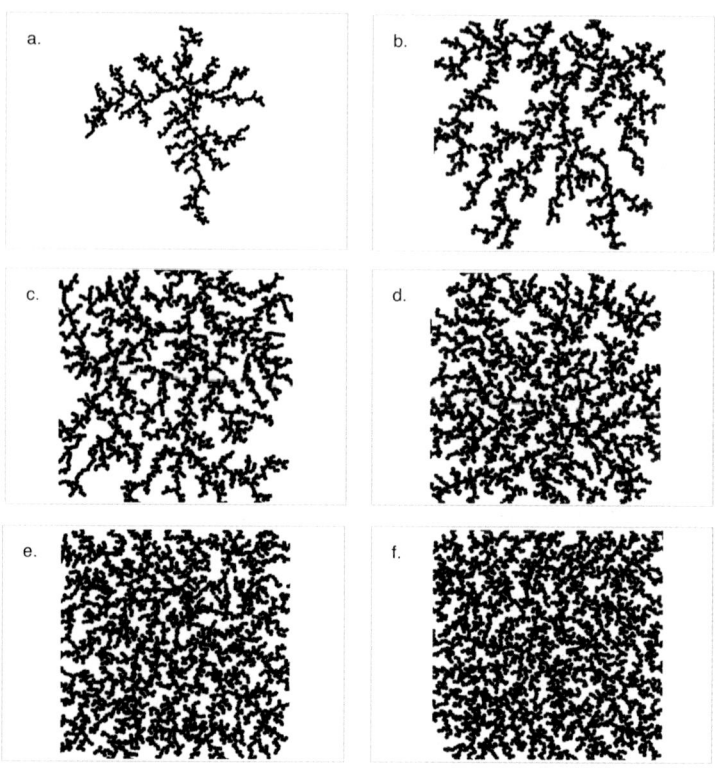

Figure 3.13 Results of DLA simulations given random starting positions and increasing the number of starting cells for each iteration.

The *information for free* that comes from emergence is critical to living systems. In fact, life itself can be seen as an emergent property of the assembly of matter. Biological systems, however, exhibit emergent properties which are much more complex than described in my DLA example, in part because they exhibit many layers of assembly and hence many layers of emergence.

SELF-ASSEMBLY AND EXTERNAL CONTROL

Biological systems are *self-organizing*, and biological construction is a process of *self-assembly*. In contrast, in human manufacture, we act as the agents of assembly, or we assemble other agents (machines) which, themselves, direct assembly. But this distinction can be unhelpful. Using, for example, Zhang's definition of biological self-assembly as "the spontaneous association of numerous individual entities into a coherent organization and well defined structures to maximize the benefit of the individual without external instruction" (Zhang, Yokoi and Zhao, 2006: 231), a key to understanding self-assembly is to define the boundary between the *world* and the *self* such that external interference can be distinguished from internal control. However, to use this definition, we need to make a clear distinction between the *self* being assembled and instructions which are *external* to it. In my examples described previously, simple molecular assembly is distinct from the assembly of biological molecules, which, in turn, is distinct from the assembly of multicellular systems. However, each level within our system is actually dependent on the layers below.

My examples are not only simplified, they are reductive. Through a reductive understanding of biological systems, it is possible to find these discrete examples of self-assembly, but only by ignoring the porosity of their boundaries. For example, the production of a protein in a living cell is a masterpiece of self-assembly when seen discretely. Protein molecules create a huge range of different types of material in many different phases, from fluids and gels to solids (McManus *et al.*, 2016). While complex, each protein is composed of combinations of only 20 different parts, which are molecules known as amino acids. Proteins achieve their functionality from their complex forms, which are assembled because the amino acids are sequenced in such a way that the long chain of molecules folds into specific patterns, leading to a geometry which, given the same sequence of amino acids, will almost always result in a predictable final form. This is a quintessential example of self-assembly, *without external instruction*. However, it could be argued that the amino acid sequence requires the external instruction given by DNA and the process of transcription and translation which converts information held in the DNA sequence into the amino acid chains in the first place. The protein must also 'know' to be made. In biological terms, proteins are only produced when they are needed by the cell. They are thus regulated, and this is also a form of external instruction. Set our boundaries to include the process of translation and protein regulation, and we now have a new and more complex multilayered form of self-organization.

We can play this game indefinitely. Where do we draw the boundary of the *self* in self-assembly?

BEYOND THE 'SELF'

This game is a mischievous one. By continually pushing the boundaries of what we consider a 'self' in biological systems, a case could be made that all assembly is self-assembly. As our bookcase-loving alien species sees us at the beginning of the chapter, the biosphere of earth is one continually self-assembling system with no distinction between geological formations, trees and houses or flatpack furniture.

Challenging the boundary conditions of these self-assembling layers, much like challenging the distinction between the *inside* and the *outside* of biological systems (as I did in Chapter 2), is useful, as it allows a different view of the systems we are studying. In the context of biology, we cannot conceive of a system which is disassociated from *external instruction*. Living cells survive through a constant flow of energy and information from the outside world, and so *self*-organization and *self*-assembly are constantly modified by the conditions the *self* finds itself in. These flows are very important to the process of design with biology – they provide us, as I will show, the parameters within which we can design. Armstrong, for example, suggests that the processes by which biological systems build are based on radically different sets of 'design' logics from human manufacturing. Referencing the work of Prigogine (Prigogine and Stengers, 2018), Armstrong argues that, unlike industrial systems, which are founded on the principles of Newtonian physics and the dynamics of *closed* systems seeking equilibrium, biological systems are defined by being far out of equilibrium. Biological systems are thus thermodynamically *open*, being fed by and feeding out energy (Armstrong, 2015).

If we take the 'self' out of assembly, a different set of questions arises relating to how assembly occurs in biological systems. What are the ingredients for assembly, and, for the designer, what do we instruct in order to construct?

CONCLUSION

What I have attempted to show here is that in any biological assembly system, the answer to the question of where the information is will depend on the scale. Information resides in either:

- the parts themselves (as in molecular assembly),
- the parts themselves alongside templates which act as information holders (as in the assembly of biological molecules) and/or
- the environment which is patterned through the interaction between the parts (as in multicellular assembly).

As we assemble many simple parts, which may each exhibit simple rules of patterning, we see emergent behaviour. The information content increases as the system develops, but information is not present before the parts assemble. Emergence can occur across many scales, but within biological systems, it is prevalent simply because so many parts are involved, and emergence occurs across as well as within scales.

Our scientific disciplines have developed tools of both visualization and control associated with each scale. In chemistry, the relative concentrations of parts, the amount of energy in the environment and the bounding of space can be used to control chemical reactions. In molecular biology, we edit the template molecule of DNA to alter mRNA and the subsequent assembly of proteins. In tissue engineering, multicell morphogenesis can be controlled through the shaping of the physical environment on which the cells are grown. However, considering these layers independently is limiting.

The idea of self-assembly, I have suggested, depends on us finding parts which assemble without the need for external instruction. But the idea of 'self' is dependent on us setting sometimes arbitrary boundaries between one scale of assembly and another. Asking where the information resides is challenged by causalities across scales of assembly, and this will be explored further in the next chapter.

REFERENCES
Armstrong, R. (2015) *Vibrant Architecture: Matter as Codesigner of Living Structures.* Warsaw: De Gryter Open.
Ben-Jacob, E. (1997) 'From Snowflake Formation to Growth of Bacterial Colonies II: Cooperative Formation of Complex Colonial Patterns', *Contemporary Physics*, 38(3).
Brown, R. (2013) *Extreme Tissue Engineering: Concepts and Strategies for Tissue Fabrication.* Oxford: Wiley-Blackwell.
Davies, J. A. (2005) *Mechanisms of Morphogenesis: The Creation of Biological Form.* London: Elsevier.

Davies, J. A. (2008) 'Synthetic Morphology: Prospects for Engineered, Self-Constructing Anatomies', *Journal of anatomy*, 212(6), pp. 707–719. https://doi.org/10.1111/j.1469-7580.2008.00896.x.

Davies, J. A. (2016) 'Machines for Living in: Connections and Contrasts between Designed Architecture and the Development of Living Forms', *Architectural Research Quarterly*, pp. 45–50. https://doi.org/10.1017/S1359135516000154.

Dill, K. A. and MacCallum, J. L. (2012) 'The Protein-Folding Problem, 50 Years On', *Science*, 338(6110), pp. 1042–1046. https://doi.org/10.1126/science.1219021.

Gibney, B. R., Rabanal, F. and Dutton, P. L. (1997) 'Synthesis of Novel Proteins', *Current Opinion in Chemical Biology*, 1(4), pp. 537–542. https://doi.org/10.1016/S1367-5931(97)80050-6.

Gibson, D. G. *et al.* (2008) 'Complete Chemical Synthesis, Assembly, and Cloning of a Mycoplasma Genitalium Genome', *Science (New York, N.Y.)*, 319(5867), pp. 1215–1220. https://doi.org/10.1126/science.1151721.

Haeckel, E. (2012) *Art Forms in Nature*. Munich: Prestel Verlag.

Holland, J. (2014) *Complexity: A Very Short Introduction*. Oxford: Oxford University Press.

Kühner, S. *et al.* (2009) 'Proteome Organization in a Genome-Reduced Bacterium', *Science*, 326(5957), pp. 1235–1240. https://doi.org/10.1126/science.1176343.

McLain, S. (2017) *The Physics of Life: How Water Folds Proteins*. Available at: www.youtube.com/watch?v=SaSntOkK-Vk (Accessed: 21 September 2018).

McManus, J. J. *et al.* (2016) 'The Physics of Protein Self-Assembly', *Current Opinion in Colloid and Interface Science*, 22, pp. 73–79. https://doi.org/10.1016/j.cocis.2016.02.011.

Norn, C. H. and André, I. (2016) 'Computational Design of Protein Self-Assembly', *Current Opinion in Structural Biology*, 39, pp. 39–45. https://doi.org/10.1016/j.sbi.2016.04.002.

Papadopoulou, A., Laucks, J. and Tibbits, S. (2017) 'From Self-Assembly to Evolutionary Structures', *Architectural Design*, 87(4), pp. 28–37.

Perez-Gomez, A. and Pelletier, L. (2000) *Architectral Representation and the Perspective Hinge*. Cambridge, MA: MIT Press.

Philips, R. *et al.* (2012) *Physical Biology of the Cell*. New York: Garland Science.

Prigogine, I. and Stengers, I. (2018) *Order out of Chaos: Man's New Dialogue with Nature*. London: Verso.

Shapiro, J. A. (1998) 'Thinking about Bacterial Populations as Multicellular Organisms', *Annual Review of Microbiology*, 52, pp. 81–104. https://doi.org/10.1146/annurev.micro.52.1.81.

Turing, A. M. (1952) 'The Chemical Basis of Morphogenesis', *Philosophical Transactions of the Royal Society of London*, 237(641), pp. 37–72.

von Goethe, J. W. (2009) *The Metamophosis of Plants*. Cambridge, MA: MIT Press.

Zhang, S., Yokoi, H. and Zhao, X. (2006) 'Molecular Design of Biological and Nano-Materials', in Bar-Cohen, Y. (ed.) *Biomimetics: Biologically Inspired Technologies*. London: Taylor and Francis.

Fabrication in the living
Chapter 4

FROM ASSEMBLY TO FABRICATION

In Chapter 3, I outlined the logics of assembly for biological systems at different scales. I showed how, from molecules to cells, biological systems assemble by means of information held in the parts themselves, in their environment and in templates which separate information from the parts being assembled. As designers, we need to understand these sources of information because, by changing them, we can change the process of assembly. I also introduced the idea of 'information for free' through patterns which are an emergent phenomenon based on the interaction of many simple parts but where no single source of information exists. Emergence happens at many scales in biological systems, and, as a result, it is not possible to ascribe biological assembly, as a whole, to individual sources of information. There are no blueprints for living things. I have also critiqued the idea of self-assembly. Self-assembly, I have suggested, implies the definition of a *self* which is independent from *external* control. In reality, no clear boundaries between the self and external influences or, as discussed in Chapter 2, between inside and outside exist for biological systems. While, as I did in Chapter 3, we can view assembly in biological systems at a series of scales, in practice, these scales cannot be easily separated.

This chapter will build on these ideas by moving beyond *assembly* to *fabrication* which, as I will describe it, involves an understanding of causalities across scales. This description will be illustrated with selected examples, and this chapter will attempt to outline different design frameworks and their weaknesses, introducing ideas of top-down and bottom-up design, *in vitro* and *in vivo* processes, parameterization and decomposable systems and will conclude with an adaptation of Waddington's notion of a 'necessary path' of biological development or 'creode'. None of these frameworks is perfect, but they do give us different insights, and I will illustrate them with examples of biological fabrication projects.

As I describe in Chapter 2, synthetic biology borrows metaphors from fields such as mechanical and electrical engineering, and the design of modern engineered systems is so wrapped up in descriptions of parts and hierarchies that it is difficult to move past these ideas as a framework. In fact, scientific fields are formed around such hierarchies. Henriques, for example (2003), in his 'tree of knowledge' system, divides scientific knowledge into scales and complexities of matter. His 'tree' is made up of tiers – each representing an organization of matter from atoms to molecules, growing in complexity to include life, the mind and culture. After the big bang occurred (some 15 billion years ago), matter formed, and this is now studied in the physical sciences with highly complex forms of matter (life) evolving 700–800,000,000 years ago now the subject of biology, and through the evolution of humans, the study of the most complex organization of matter (the mind) becomes the study of psychology (Figure 4.1).

This tree of knowledge could be seen as a tree of assembly. While each tier in the tree is built on the one below, each of the disciplinary groups is focused on discrete theories and methods which, according to Henriques, are associated with different levels of complexity.

Central to my argument is the notion that, while useful as an analytical framework, the hierarchy of parts within parts which I described in Chapter 3 falls down when we come to intervening with biological systems as a form of fabrication. In Figure 4.2, I have adapted a description of the hierarchical assembly of hair from Vincent (2012). Hair can be drawn as a hierarchy of parts, starting with the amino acids which assemble to form helix-shaped molecules of keratin wound into fibres, making up macrofibrils formed by cells. While this description of parts is useful in

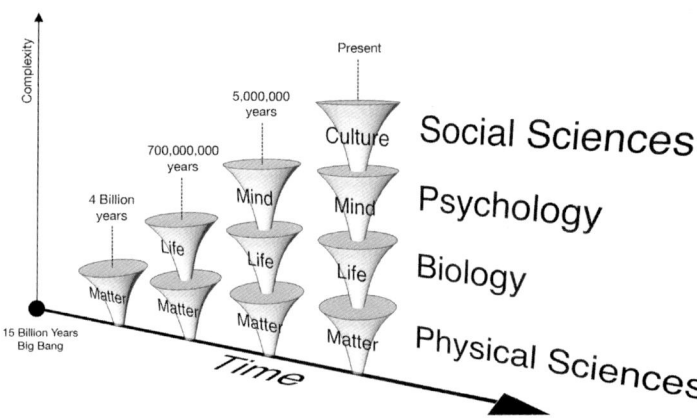

Figure 4.1 The 'tree of knowledge' system based on a diagram by Greg Henriques (2003), showing the emergence of matter since the big bang into increasingly complex forms.

Fabrication in the living

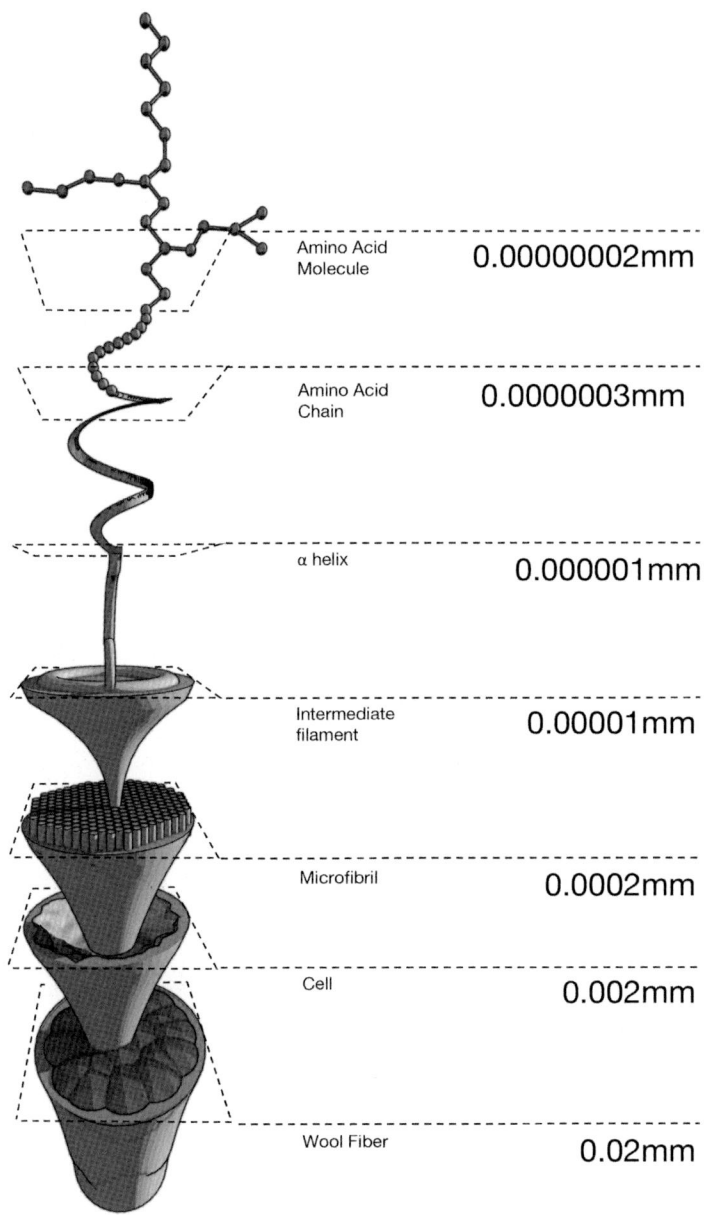

Figure 4.2 The assembly of hair at multiple scales, adapted from Vincent (2012).

understanding the composition and structure of hair, we cannot simply reverse-engineer a hair by only understanding the parts which make up the whole. We would need to bring together many disciplines to understand the detail of each level of assembly, including biochemistry, molecular biology and cell biology, to explain the different scales of assembly. Understanding the structure of amino acids, if we consider them the fundamental building blocks of hair, tells us as much about the structure of hair as a brick tells us about the structure of a house. Furthermore, we need to know how the levels are connected. Starting with amino acids alone, we would be faced with the same problem as trying to reassemble the DLA pattern described in Chapter 2 by reference to the end result. As Gilbert *et al.* point out, we would miss the interdependencies of the different levels of assembly:

> when we try to explain how the whole system behaves, we have to talk about the context of the whole and cannot get away talking only about the parts. . . . This is not to say that each level is independent of the lower one. To the contrary, laws at a level may be almost deterministically dependent on those at lower levels; but they may also be dependent on levels "above."
>
> (Gilbert *et al.*, 2000: 3)

Because of the interdependency of levels, I will show in this chapter that we need to integrate these different levels when thinking about the design of biological fabrication.

TOP DOWN V. BOTTOM UP

If we were to accept that biological fabrication is framed as a hierarchy of scales, we need to ask at what level in the hierarchy we intervene and in what direction. These questions are sometimes framed as a distinction between top down and bottom up. Notions of top down and bottom up have a range of definitions in design, but, in synthetic biology, bottom-up design is seen in attempts to construct novel artificial life from scratch. Work on protocells, for example, looks to develop functioning cells from basic chemical components, including using self-assembling lipids to create cell membranes which act as containers for controlled chemical reactions (Stano and Mavelli, 2015). In contrast, top-down design modifies existing organisms and introduces novel metabolic pathways or signalling systems (Roberts *et al.*, 2013).

In the field of engineered living materials (ELMs), an aspiration is to be able to achieve material structures which are self-assembled through cell factories capable of organizing matter into complex structural forms. The example described by Nguyen *et al.* (2018) is of building a synthetic system which would work in the same way as a tree – starting with a seed 'which carries the essential information needed to create the tree' and which, given appropriate energy and carbon sources, fabricates materials capable of self-maintenance and renewal. While seeds do exist for the fabrication of trees, there are no seeds yet for the fabrication of buildings. Truly bottom-up design is at a very early stage, and we are currently unable to create life from scratch and, in biological fabrication, it is not yet possible to synthesize substantial materials or structures using protocells. In reality, therefore, when we discuss bottom up versus top down, we are usually making a reference to the complexity of the organism we are working with and the degree of influence we have in defining the outcome of a fabrication process. We can illustrate this with reference to the fabrication of cellulose-based materials.

IN VIVO AND *IN VITRO* BIOLOGICAL FABRICATION

Cellulose is one of the most common biologically synthesized materials on earth (Ross, Mayer and Benziman, 1991), which in part is due to the development of the highly controlled extracellular synthesis of cellulose in plants (Cosgrove, 2005). The controlled synthesis of cellulose into hierarchical fibre structures and the addition of additives such as lignin create cellulose composites which are more or less hydrophobic and thus structurally stiff (when water absorption is low) or rubbery (when water absorption is high). The composition of cellulose in plants can also change over time from, for example, the flexible cellulose stem of a new sapling to the tough drier material constituting the superstructure of a tree trunk. A wide range of organisms, including bacteria, are able to produce cellulose materials, and the diverse properties of cellulose and its relative abundance make it an interesting case for biological fabrication.

Modern cellulose-based materials are derived from natural sources, but, in all cases, there is a degree of post-production. Our design processes are adapted to make use of this natural material, sometimes in a relatively untreated form, as in the case of using green oak as a construction material. The tree must be cut down, however, and shaped into constructible parts. In an industrial society, the need for a greater range of materials has given rise to a greater number of post-production processes, for example, the manufacture of paper, which involves wood being pulped, bleached and reconstituted into thin sheets which are totally transformed from their original form. We also tend to simplify material structure in post-production. The crude fibrous structure of paper retains none of the hierarchical complexity of wood. Post-production enables a greater range of materials for different uses which can be precisely tuned, but these industrial post-production processes are often energy intensive and require substantial infrastructure.

An alternative approach would be to work with cellulose as it is being produced by the organism. This means intervening in the process of material construction *in vivo*, which translates to *in the living*, through means of *in vitro* control. *In vitro* literally translates to *in the glass* but here refers to a broader notion of the human control of the chemical and physical environment. I will refer to *in vitro* and *in vivo* here as *domains of information*, where *in vivo* is information contained within a cell and *in vitro* is information contained in the environment. Fabrication results from the interaction between *in vivo* and *in vitro*. The nature of this intervention depends on the organisms involved, the tools we are using and the desired outcome.

We could start with a tree growing in the wild. The tree has been assembled without human intervention, that is, entirely *in vivo*. The cellulose lignin composite (wood) can still be harvested, but we must adjust our design outcomes to suit this 'found' material.

Planting trees in a managed forest environment could be seen as intervening in the manufacture of the trees through, for example, changing the chemical composition of the soil to enhance growth and choosing planting patterns and species types to optimize yields. Here we are introducing *in vitro* interventions to alter an *in vivo* process.

Go further and we begin to directly intervene in the tree's morphology through, for example, constraining its development. Bonsai trees, which are grown in nutrient-poor soils, have their branches 'strangled' and are regularly pruned (Korner, Pelaez and John, 1989), are an example of this. Or the living root bridges in India (Chaudhuri, Bhattacharyya and Samal, 2016), in which roots are grown over many years, then tied and intertwined to create suspension bridges. Here humans have directed growth with specific morphologies in mind by constraining and guiding growth to achieve a desired material, albeit without precise control of the resulting forms and material properties.

Each process represents a greater level of control, but each could be seen to represent a top-down process. In all cases, we are intervening in a process which is already happening. Wood retains its complexity so that the basic material parts and their cellular organization would be fundamentally the same. We are intervening to change only the morphology of the tree.

Conversely, a bottom-up method could start with simple (compared to wood) bacterial cellulose. Bacteria are not capable of the same multiscale assembly of cellulose as plant cells. Growing a sheet of cellulose (Figure 4.3) requires *in vitro* intervention to alter the *in vivo* process.

Pure cultures of bacteria need to be grown in a controlled chemical and physical environment. Although bacterial cellulose has a much simpler structure than wood, through this *in vitro* control, we have a greater potential to influence the outcomes. While we can influence the morphology of the wood from a tree, we can influence more fundamental material properties in cellulose through the addition of additives, with genetic changes to the bacteria cells or by altering the growth conditions (Shoda and Sugano, 2005). This

 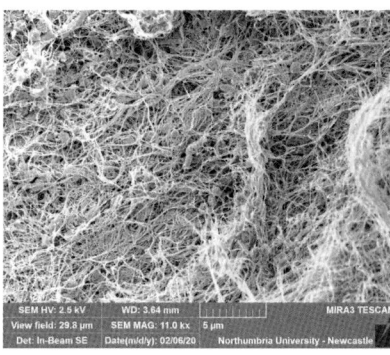

Figure 4.3 A sheet of bacteria-grown cellulose grown in a flask with electron microscope images showing the structure of the material.
Source: Images courtesy of Sunbin Lee and Yunhong Jiang.

bottom-up process starts with a relatively simple process of bacterial cellulose production and adds complexity to the fabrication process and material outcome.

In practice, we often need many points of intervention across different scales involving both *in vivo* and *in vitro* interventions. Zolotovsky's PhD thesis on guided growth offers one of the most comprehensive examples of biological fabrication using bacterial cellulose and involves the use of synthetic biology and the orchestration of *in vitro* and *in vivo* processes. A full description can be found in her PhD thesis (Zolotovsky, 2017) and subsequent publications (Zolotovsky, Gazit and Oritz, 2017, 2018). Zolotovsky proposes a design framework based on different scales of intervention which, at first sight, seems to follow the scale hierarchy described previously.

At the *nano scale*, Zolotovsky proposes the use of tools from molecular biology to modify cells to create a tunable patterning system. The design, which is based on the work of Basu *et al.* (2005), is composed of a 'lawn' of two types of cell: senders and receivers. The sender cells release signalling molecules which operate across the field of cells. Signalling molecules are detected by the receiver cells which, in turn, are responsible for making a product. In Basu *et al.*'s paper, the receiver cells are programmed to produce different levels of green fluorescent protein (GFP) depending on the concentration of the signalling molecules received from the sender. The cells are tuned to respond only to certain concentrations of the signalling molecule. If the sender cells are placed in the middle of an agar plate, for example, their chemical signal will diffuse, creating a chemical gradient where the highest concentrations will be found next to the sender cell colony, dropping away with distance. The receiver cells are inhibited from producing GFP if the concentration of the signalling molecule is too high or too low, so a ring appears around the signal cell colony as receiver cells at the distance corresponding with the right levels of concentration for GFP production are illuminated under UV light. In Zolotovsky's example, the same system is proposed, but rather than GFP, the receiver cells are designed to produce a fusion protein which increases the number of crosslinks between the chains of cellulose molecules, and this selectively increases the density of the cellulose film (Figure 4.4). Using such a technique, it would be possible to create a

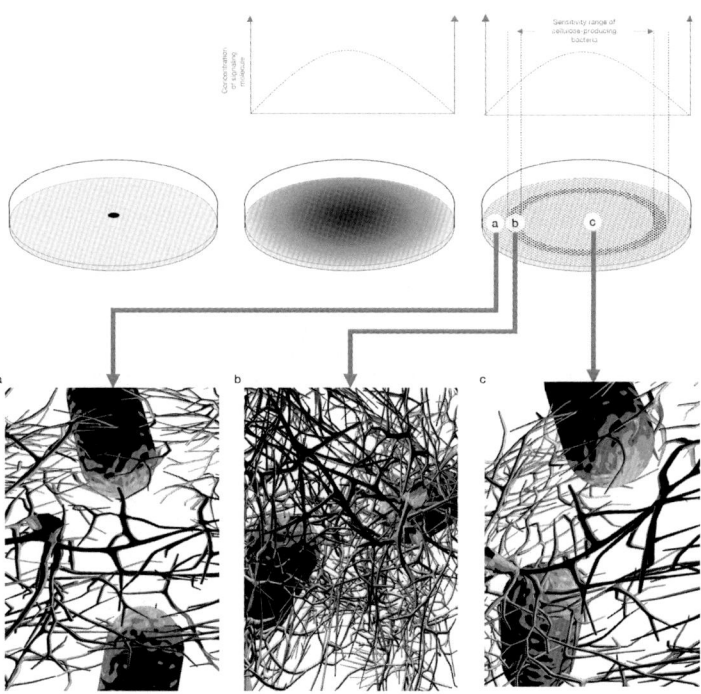

Figure 4.4 Diagram based on the work of Katia Zolotovsky to show the patterning of bacterial cellulose using a synthetic biology system in which the bacterial are tuned to respond to a chemical signal by increasing the destiny of crosslinks in response to a specific chemical signal strength.

graded material whose physical properties differ in different places, becoming denser in some places and less dense in others.

At the *meso scale*, Zolotovsky experiments with a number of fabrication methods which aim to alter the growing environment and 'post-production' of the cellulose and 'bridge the gap' between the nano-scale assembly described previously and the macro scale of the final parts. These methods included agitating the growing cells, adding other chemicals such as polyvinyl acetate (PVA) and magnetite to make composites and treatments including freeze drying and air drying as well as techniques for moulding.

At the *macro scale*, Zolotovsky proposes a 'macro-fluidic pneumatic interface', which is essentially an advanced irrigation system or scaffold, which is computer programmed to enable different growth conditions through the controlled feeding of the cellulose cultures and testing the basic physical properties of the materials which result (Figure 4.5). This plumbing acts rather like a vascular system, feeding the cells and regulating their growth and the production of cellulose. This macro-scale intervention

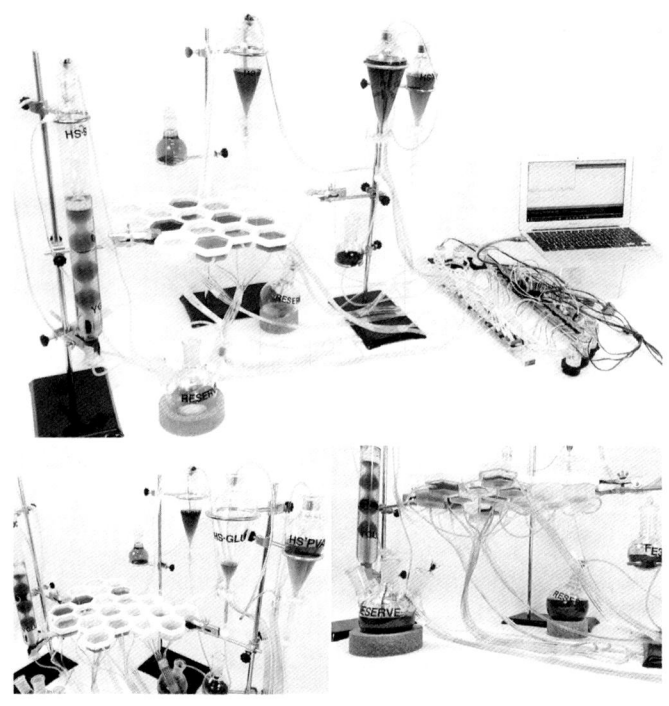

Figure 4.5 The 'macro-fluidic pneumatic interface' for the cultivation of bacterial cellulose.
Source: Image courtesy of Katia Zolotovsky, Merav Gazit and the Ortiz Laboratory for Structural Biological Materials at MIT.

also introduces a new domain of information: *in silico*, meaning information held within a computer, which is, in turn, altering the *in vitro* parameters of the fabrication process.

The scale hierarchy used here, while a useful structuring device for the thesis, does not necessarily distinguish between the interventions and the effects. For example, Zolotovsky uses the tools of molecular biology to propose the nano-scale genetic engineering of cells to develop the patterning system which works at both the scale of molecules cross-linking cellulose chains and the pattern of rings which form at the macro scale and would be visible to the naked eye. At the same time, the macro-scale construction of scaffolds and macro-fluidic devices regulate the growth of micro-scale bacteria through the provisions of nutrients, which result in changes to the nano-assembly of the cellulose itself.

In attempting to 'guide the growth' of bacterial cellulose through *in vivo*, *in vitro* and *in silico* interventions, Zolotovsky is working with causalities and effects which cross scales, both bottom up and top down, which makes it impossible to consider altering the assembly of any part of the cellulose material without referencing other scales of assembly. We have found countless examples of these *vertical causalities* in our own work.

BOTTOM-UP CAUSALITY: THINKING SOILS
In the Thinking Soils project, we envisaged a material which would, in response to loads placed on it, remodel itself, strengthening where additional support was required. The project was inspired by the way in which bone remodels under repeated load, strengthening where it is required by increasing the density of bone matter.

We proposed the development of a material embedded in bacterial cells which detects forces applied to them and synthesizes strengthening materials in response. In other words, this living material would compute and respond to its environment. We envisaged, for example, a construction situation in which engineered bacterial cells could be seeded in the ground beneath a construction site and, rather than digging foundation trenches and filling them with reinforced concrete, the soil would respond to loading by triggering a process of biomineralization, strengthening the ground and resisting the increasing load. We are (at the time of writing) still some distance from this aim, but our initial experiments and simulations begin to reveal the complexity of the systems we are dealing with and the challenge of designing a system of assembly at multiple scales simultaneously.

As with Zolotovsky's Guided Growth project, we started with an approach based on the separation of the design problem into multiple scales. In our case, we built a collaborative team based on separate disciplines, making a distinction between the synthetic biology work (requiring nano-scale manipulation of genes), the microbiology work (associated with the study of individual bacteria cells in their context) and the civil engineering 'macro' work (associated with measuring and modelling the forces through volumes of soil). However, by integrating these different scales, we began to identify a broader set of interactions between parts of our system across scales.

A starting point was to seek to understand how a bacterial cell might become a pressure sensor. In Chapter 3, I described the basic principles behind the central dogma, with genes being transcribed to RNA molecules which are in turn translated into chains of amino acids which fold to form proteins. Proteins go on to enable the cell's metabolism as well as building the cell's structure. An important factor in this process of transcription and translation, as I described in Chapter 3, is that both processes are regulated. Most proteins within a cell are not continually expressed but are expressed in ways which are context dependent. In the language of synthetic biology, gene circuits have regulatory switches; if we know what factor turns them on and off, we can, in theory, reengineer this process such that we have a switch which is under the control of the regulatory factor we are interested in, that is, pressure in the Thinking Soils case.

Using a process called RNA sequencing, we were able to take a snapshot of the genes being expressed in a bacterial cell when the cell is at normal atmospheric pressures and again when the cell becomes pressurized. While our results are far from conclusive, we identified candidate genes which showed significant changes in expression between our pressurized and unpressurized conditions and engineered the DNA to make the cells fluoresce under UV light as they were put under pressure. We could then assess the genetic response of our cells by measuring the amount of fluorescence at different pressures. A more in-depth study of our research can be seen in Guyet *et al.* (2018) and Dade-Robertson *et al.* (2016).

Imagine if, rather than fluorescing, the cells were engineered to synthesize a material which strengthens soil. In this proposed scenario, we would want to see a linear relationship between pressure and the increased expression of our gene of interest. Under unpressurized conditions, our gene of interest would be dormant, but as pressure increased, it would be progressively turned up, making more strengthening material. However, this is not what we see in our experiments.

Gene expression rarely has a simple linear relationship between the input and the output. When we performed our pressure experiments, it was with the expectation that the bacteria would respond to elevated water pressure in a saturated soil which had been loaded. As a saturated soil is loaded, water pressure will build up temporarily before the water dissipates and the soil sinks (consolidates). In our sample bacteria, over 100 genes were shown to have a response to increased water pressure and, after selecting the most likely candidates, we were able to engineer the bacteria so we could evaluate the response of the 'pressure-sensing gene' to a range of pressures. This experiment and an analysis of the results demonstrated two important things about this pressure response.

First, the bacteria were not responding to the pressure directly but rather indirectly – most likely as a response to increased levels of oxygen, which created stresses in the cell (Guyet *et al.*, 2018). This means that our pressure sensor may actually be an oxygen sensor (raising questions about its usefulness in an environment like a soil where oxygen levels may vary).

The second insight was that the response of the 'pressure sensor' was not linear. The pressure sensor is on whether or not the bacteria are pressurized. As we increase pressure, the expressed gene goes up to a point and then flattens before rising again. This sort of non-linear response is not unusual. We see these sorts of responses in response to a stimulus as well, with Sato *et al.* (1996) and Ishii *et al.* (2005) showing pressure-regulated genes (under the control of gene 'switches' or what are described as promoters) which seem to operate at a very specific range of pressures.

Notwithstanding the issue of unpredictability and, given a relatively simplistic understanding of our cell and the characteristics of pressure-sensing genes, we can still begin to predict larger-scale effects through a computational model. Figure 4.6 shows a 3D grid

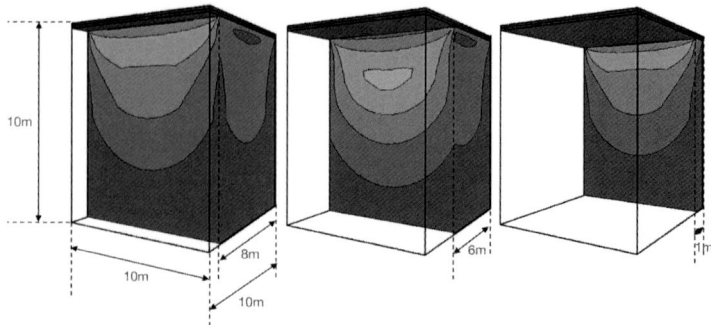

Figure 4.6 Computer visualization of a 3D volume of soil loaded from the top showing the distribution of forces through the soil. The lighter colours indicate the greater forces. The cube has been cut at points through the cube to show cross-sections through the soil volume.

representing a 10 × 10-m volume of homogenous 'soil' containing our engineered bacterial. The cube is divided into smaller areas for analysis. In the simulation, we can measure the forces through the soil when the cube is evenly loaded from the top. The computer model performs a type of finite-element analysis with the centre of each cube representing a point of analysis (see (Dade-Robertson et al., 2016, for more information). We can visualize the results of this analysis by cutting a slice through the soil and colour coding (in this case in greyscale) the magnitude of the force at each point.

We can then map the results of our gene expression data. In this case, we can draw a box at each analysis point in the soil which is proportionate to the level of gene expression. Very crudely, if we imagine that gene expression equates to the amount of a strengthening material produced by the trillions of bacteria in each part of the soil volume, then the resulting map gives us an idea of the amount of cementation in different parts of the soil. Given our non-linear promoter response, we would expect to see bands of high levels of cementation in the soil associated with bands of force.

In Figure 4.7, we play with this idea using hypothetical promoter responses with two different profiles. The image shows a slice through our soil volume with cubes drawn at a size representing the amount of cementation we might expect to see at different positions within the soil. Figure 4.7a shows the results of bacteria with a linear

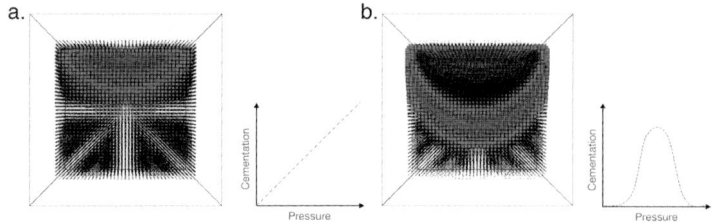

Figure 4.7 Results of a computer simulation to show the amount of cementation through a volume of soil given two different genetic pressure responses. The visualizations represent sections taken through the middle of a 10-m3 volume of soil. Cubes are drawn, the sizes of which represent the amount of cementation we might expect to see at specific points within the soil given the distribution of load and the sensitivities of the bacteria. Simulation (a) shows a linear response where the gene switch becomes more active as loads increase. Simulation (b) shows the results of a gene which is sensitive to a narrow range of pressures.

genetic response to increased pressure. We would, as expected, see the highest levels of cementation occurring where the highest pressures are felt. Figure 4.7b shows the response of bacteria with a sensitivity to a specific pressure range. In this case, we see a band of cementation.

Figure 4.8 extends this idea by using what is, in synthetic biology, referred to as a two-component system. In this example, we have two types of bacteria, a sensor and a builder. In response to elevated pressures, the sensor releases a signalling molecule which is picked up by the builder bacterial. The builder, in response to the signalling molecule, then synthesizes a cementing material. In the case of the sensing bacteria, the sensor is going to have a genetic response to force. These will vary, as described previously. The building bacteria will have a variable sensitivity to the signalling molecule produced by the sensing bacteria, and this is represented as a second gene expression curve. The interaction of these two systems begins to create interesting patterns of cementation within the soil as one system begins to disrupt the other, reinforcing or, in some cases, cancelling the other out.

This is a simplified example and doesn't consider many of the additional factors which would affect such a system, including, for example, the fact that soil is not homogenous, and bacteria would not distribute themselves evenly. It does, however, illustrate an important

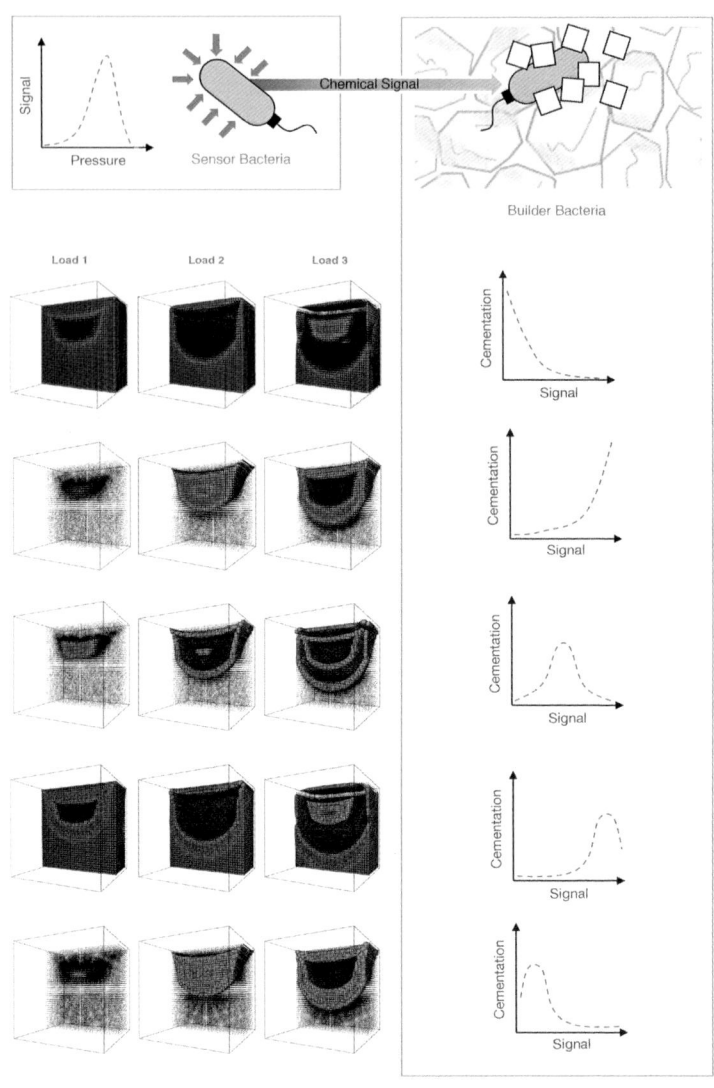

Figure 4.8 Visualizations of cementation levels within a 10 × 10-metre volume of soil containing pressure-sensing bacteria signalling to 'builder bacteria' responsible for synthesizing a cement-like material. The visualization shows different cementation patterns given various combinations of sensitivities in the builder bacteria's response to the signalling molecule.

point. These changes, represented by each of the simulations, constitute nano-scale changes in the genetic code but result in changes to the cementation of soil volumes over a number of meters.

TOP-DOWN CAUSALITY: BIO-INDUCED MINERAL CRYSTALS

In the Thinking Soils example, I illustrated bottom-up causality. In my final example, I want to introduce a top-down causality using a simple experiment conducted to grow mineral crystals. A classic and often-cited example of natural biological fabrication is the construction of a shell. Abalone shells, for example, while composed of 90% calcium carbonate, manage to achieve tensile strengths 3000× greater than in non–biologically synthesized calcium carbonate through the addition of proteins and the highly controlled synthesis of crystals in a series of tiles of the mineral aragonite, known as nacre (Lin, Chen and Meyers, 2008). The synthesis of nacre is of great interest to material science, as the prospect of being able to construct a synthetic shell could lead to a new generation of functional materials with similar properties to ceramics but which are synthesized at room temperatures and pressures. In previous work, we have used the example of the construction of a synthetic shell to indicate the challenges of biological fabrication (Dade-Robertson, Ramirez-Figueroa and Zhang, 2015). While the structure of nacre in abalone shells has been well studied, challenges in observing nacre growth live mean that there are still gaps in our knowledge as to how it is fabricated. However, we do have some insight into the factors which govern nacre growth and structure and which enable the complex structures we see in this functional material (Cartwright and Checa, 2007).

In an abalone shell, the epithelial surface of the mantle creates a layer which chemically accelerates the process of mineralization by sequestering ions from seawater. The abalone is able to initiate this self-assembly process by altering the surface chemistry of the mantle and producing proteins which are excreted outside its cell, providing points of crystal nucleation, biasing certain axes of crystal growth and templating the crystal formation. The abalone organism itself also acts as a template, and the continuous movement of the soft body of the creature acts to abrasively smooth out the inner surface of the shell. Changes in the environment also play their part, with layers of calcium carbonate exhibiting seasonal growth patterns (much like tree rings) and this variable growth also adding strength to the final material. If we ask 'Where is the information?' in the formation of the shell, we will find it at multiple levels in the assembly of different parts, from the gene templates which provide the information for protein construction through to the seasonal changes which lead to different growth conditions. We can subdivide these different scales of assembly, looking for points at which to intervene, but this would also mean isolating specific control factors, which in practice is very difficult to do.

We can illustrate this challenge with a simple experiment. Some bacteria types can induce the formation of calcium carbonate crystals by altering the pH of their environment and their cells help to nucleate crystals. The process of biomineralization has been of significant interest to researchers and designers, including those working on self-healing concretes (Jonkers, 2007) and biocements (Dosier, 2011). This process has also been suggested by some (Decho, 2010) as a primitive version of the processes we see in higher organisms such as abalones, in part because, while the crystals are being induced, there is also evidence of crystal templating in, for example, biofilms (Dade-Robertson *et al.*, 2017), and distinct differences in crystal morphology created by different sorts of bacteria (Dade-Robertson, Ramirez-Figueroa and Zhang, 2015).

In a series of simple experiments conducted with our design students, we began to explore how we could influence the process of biomineralization by, initially, altering the space in which the crystals form. In our experiments, we grew bacteria colonies on circular agar plates in calcium-rich media and then compared the results with agar plates which were modified, constraining the agar to smaller channels without changing any other variables. The agar plates in these examples, it was hypothesized, would act as a mould for the crystals, and we expected to see that the microscopic crystal morphologies themselves would be unchanged. This is not, however, what we found.

Figure 4.9 shows a comparison between two agar plates. In the first example (Figure 4.9a), the bacteria were placed in the centre of the plate and allowed to grow freely across the surface. In the second plate (Figure 4.9c), the bacteria were placed into a reservoir and

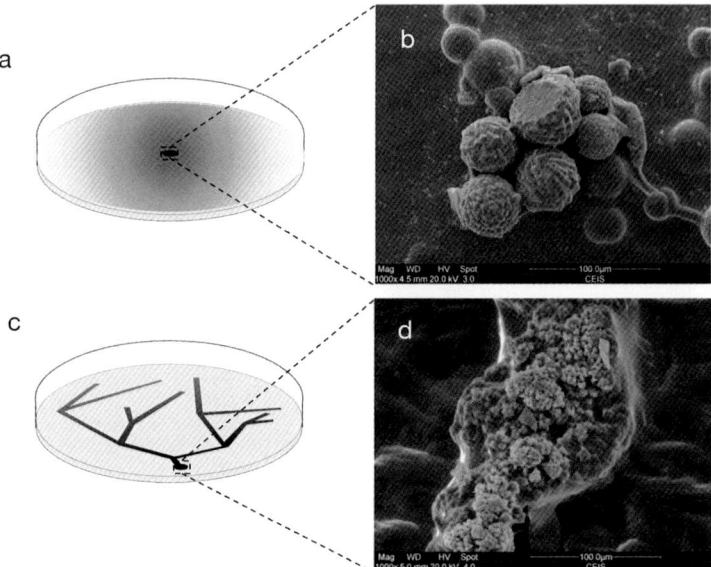

Figure 4.9 Two agar plates inoculated with biomineralizing bacteria. In (a), the plate is unmodified and in (c), the plate has been modified with channels cut so that bacterial growth is constrained to channels. (b) and (d) show electron microscope images taken of the crystals produced in each setup.

allowed to grow along the channels which contained agar of the same consistency, nutrient levels and so on. We then took electron microscope images of the crystals which formed. Figure 4.9b is an electron microscope image of the faceted spherical crystals growing on the unmodified agar plate. Figure 4.9d shows the crystals grown within the modified agar plate. The morphology of the crystals between the two plates looks very different. In setting up the experiments, we expected that the distribution of crystals would be more constrained in those plates where channels had been cut, as we were essentially casting the same materials into a different mould. We wouldn't expect the crystalline structure of concrete, for example, to be substantially different if the same mixture were poured into two different-shaped moulds. However, the electron microscope images show that altering the mould has a substantial effect on the microstructure of the crystals created by the bacteria. This is, at first sight, a surprising result; it is important to recognize that, while the channels in the second experiment are narrow, the space they create is still vast compared to the bacteria growing in them and the crystals created. A bacterial cell is approximately 2 micrometres across; the channel is about 2 mm across. For a human (approximately 1.7 m tall), this would be the equivalent of a space the size of central London. This compares to the diameter of the agar plate (approximately 10 cm), which, to the bacteria, is equivalent to roughly the size of the southeast of England. It might seem counterintuitive to consider that altering something across such massive length scales, given the size of the bacteria and the crystals they are inducing, will alter the micro-scale structure of the crystals themselves. In this case however, it does.

While a causal link is not clear, in our experiments, we observed that the agar tended to dry out more quickly in the thin channels created by the modified plates than on the larger plates. This would have impacted the bacteria's ability to grow and reproduce, as well as providing additional forces as the agar gel reduced in size. Nutrients would have been used up more quickly, and this would also have inhibited bacterial growth, as well as altering their metabolism, including their production of extracellular proteins. By altering the macro environment (in this case through a simple constraint in the geometry of the mould), we are altering multiple interrelated processes, from the metabolism of the bacteria cells to the physical constraints on the growth of the crystals.

THE CHALLENGE OF DECOMPOSABILITY
The top-down and bottom-up causalities described previously imply the need for a non-hierarchical model of biological assembly which sees events at multiple scales as affecting one another in a network of interrelationships. If we can map these relationships, then perhaps we can find places to intervene and predict cause and effect. We might, rather than drawing a tree, consider the process of biological fabrication as a network of interacting parameters. We could, following the diagrammatic language used in Christopher Alexander's *Notes on the Synthesis of Form*, propose a network diagram consisting of parameters within our biological system and relationships between them (Dade-Robertson, Ramirez-Figueroa and

Zhang, 2015). The problem with biology is that this diagram becomes very complicated very quickly.

In Chapter 2, I discussed how, in the design of traditionally engineered systems, we tend to break them into discrete and decomposable elements or modules. This means that even in network-like descriptions of biological systems, we tend to emphasize distinct hierachies composed of elements with an identifiable inside and a limited interface to an outside so that the modules can be worked on independently or swapped out easily without affecting the rest of the system. Decomposable systems have the benefit of exhibiting clusters of interactions within parts, with more limited connections between parts. These parts also often contain simpler parts which themselves are decomposable, which creates a nested hierachy. In analyzing complex biological systems, we do observe modular structures, hierachies and decomposability, and such representations are common. For example, a cell is bounded by a cell membrane, and it contains tens of thousands of macro molecules (Neidhardt, Ingraham and Schaechter, 1990) interacting through a biochemical network of approximately 1000 metabolic reactions (Ouzounis and Karp, 2000). The cell will interact with its environment, taking in nutrients and exporting waste and other extracellular materials, but the cell can be considered a bounded space with limited interfaces to the *outside*. In multicellular organisms, individual cells specialize to form discrete organs which often contain clear borders, for example, muscle, brains, lungs and so on. Thus, a cell is an individual unit at one level of the hierarchy and is part of a specialized aggregation of cells representing a discrete organ or body part.

In synthetic biology, decomposability is often considered a key requirement because, to design using a framework from traditional engineered systems, we need parts to be generalizable and interchangeable.

While it is true that decomposable or near-decomposable structures can be found in biology, biological systems differ from human-engineered systems. As discussed in Chapter 2, biological systems are thermodynamically open, as opposed to human-engineered systems, which are usually thermodynamically closed. This means that the interfaces between the inside and outside of biological parts constitute a continual flow of energy and matter without which the system would not only cease to operate, but the parts themselves may be irrevocably changed (i.e. cells would die, proteins become denatured etc.). Biological systems also have what I would describe as a transient identity. Any network pattern which describes parts and their interactions is defined by a specific context. Not all links between nodes can be considered equal, and their importance will change over time and in different conditions.

A requirement for decomposability tends to reinforce the hierachical view of biological systems and supports the idea that we can reduce and study parts independently and then build new systems from these parts. These diagrams, however, do not capture the sorts of 'vertical' top-down and bottom-up causalities I have described previously. Altering what might be represented as a single parameter can lead to systemic change. Biological systems, therefore, require us to consider networks of causality which are not easily decomposable (see, for example, Stotz and Griffiths, 2017). I want therefore to propose two new diagrams which seek to find alternative representations for biological systems and the methods we use to alter them.

DIAGRAM 1: DOMAINS OF INFORMATION
The first diagram begins to answer the 'Where is the information?' question for biological fabrication. Figure 4.10 depicts a triangle with each corner representing what I have described as the three *domains of information*: *in silico*, *in vivo* and *in vitro*. We can map different methods onto the triangle's edges, with, for example, the line between *in silico* and *in vivo* representing the computational modelling of biological systems and the synthetic synthesis (where genes are synthesized outside a cell) of genes often undertaken in synthetic biology. The *in vitro* to *in vivo* edge represents the control of the physical and chemical environment to alter the biological fabrication process, and the edge between *in vitro* and *in silico* represents the use of fabrication technologies to create scaffolds and substrates on which cells might be grown. Most of the biological fabrication processes I have described map somewhere in the middle of this triangle but with different sources of information biasing projects towards one domain or edge. Some specific examples are identified in Figure 4.10.

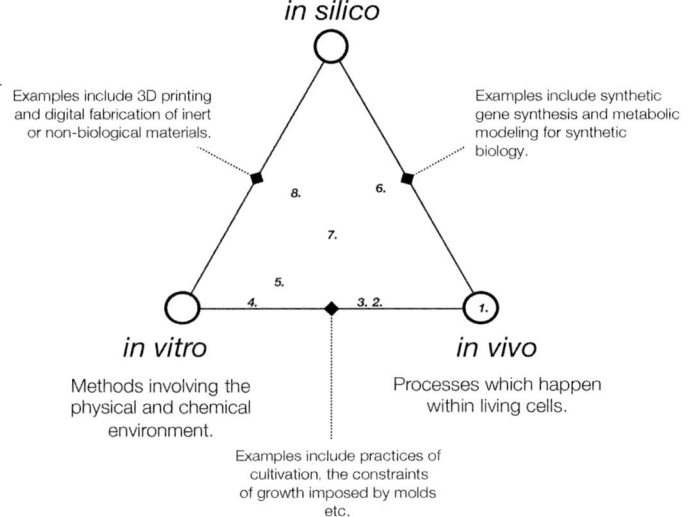

Figure 4.10 Domains of information in biological fabrication, the relationships between them and examples of biological fabrication methods with suggested mappings.

DIAGRAM 2: PHASE SPACES AND LANDSCAPES OF CHANGE

A second representation of biological fabrication may lead to a different set of design frameworks which integrate both *in vivo* and *in vitro* thinking.

In the processes of cellulose production and biomineralization described previously, we are concerned with the developmental pathways of bacterial cells – that is, their capacity to grow and subdivide in context and, in our examples, their ability to synthesize materials either by producing them within their cells and then exporting them, as is the case with bacterial cellulose, or altering their environment through the production of enzymes which lead to biomineralization. These developmental pathways are set through a combination of factors 'inside' the cell and their interaction with factors 'outside' the cell. Both processes represent the fabrication of materials which would happen 'in nature'. In both cases, however, we have co-opted these natural processes by modifying aspects of the organism's development *in vivo* and *in vitro*. The key to explaining these processes is that they are not defined by singular events but happen over time and consist of changes in phase through a flow of matter and energy. Network models describing biological systems in terms of interactions or parameters in decomposable networks don't pick up these transitions and flows.

I reflected in Chapter 2 that our metaphors often frame our understanding of a subject – giving a structure which is defined by the logic of the metaphor rather than the thing it represents. In approaching our alien technology of biological fabrication, however, metaphors offer at least a way to frame the questions we want to ask, and alternative metaphors might, therefore, offer some directions and lead to new design frameworks.

Our alternative to the hierarchy or biology-as-network model comes from the work of the developmental biologist Waddington. His seminal book *The Strategy of Genes* (2014) invokes the metaphor of a landscape based the concept of phase space. In Waddington's description, a phase space represents a multidimensional space, the dimensions of which are based on some property of the object they represent. An example is illustrated in Figure 4.11. Here the x axis represents temperature, the y axis represents size and the z axis represents time. In modelling this space, we can represent change as a trajectory through the phase space. The diagram shows our object (represented as a ball) and its movement through the phase space as it heats up and grows, with a start position and an end position and with the phase change represented as a line.

If we were to imagine modelling a biological cell in this way, we would have to represent a highly multidimensional space with many thousands, or even millions, of dimensions representing the potential cell states, including, for example, descriptions of

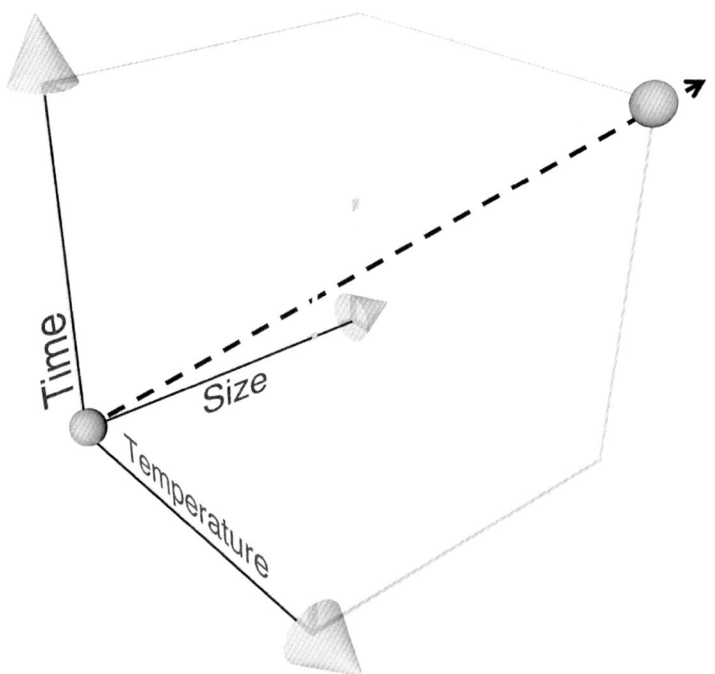

Figure 4.11 Diagram to represent a phase space where axes represent the properties of an object and a trajectory indicates the change in phase over time. Here the object, represented by a ball, increases its size and temperature over time.

biochemical interactions, the cell's morphology and so on. In cell development, however, Waddington describes developmental pathways as like canals. A cell travels through every possible position within a phase space, but instead developmental processes lead to certain pathways and end results. This 'canalization' means that cells in multicellular organisms will tend to follow certain paths through a phase space, arriving at an end point associated with the cell's specific function and specialization. Waddington draws a simplified description of the phase space as a landscape, and I have redrawn a version of this diagram as Figure 4.12a. Here the phase space model is flattened to a 3D surface representing the developmental landscape for the cell. The

a.

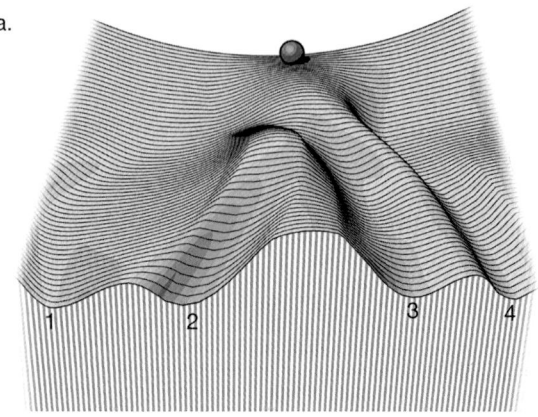

b.

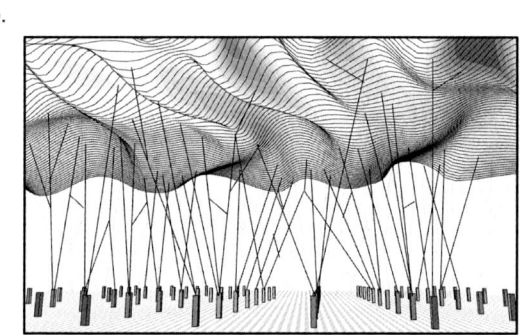

Figure 4.12 Drawings based on Waddington's epigenetic landscape (Waddington, 2014), showing an inclined landscape of valleys to describe pathways of development for a cell, which is represented by a ball. Different pathways lead to different specializations represented as destinations, indicated by numbers 1–4 in (a). (b) represents the genes (shown as stakes) and their expression is represented as 'guide ropes' which control the landscape above.

surface is inclined so that a ball, representing the cell state, placed at the top would tend to roll down. At its starting point, the cell is undifferentiated, but as the ball rolls down the landscape, it is guided into channels and will tend to roll in certain directions. The end state (at the bottom of the plane) represents a specialization, with the cell becoming, for example, part of the tissues for the eye or brain or muscle. These trajectories Waddington calls *creodes*, translated from Greek as the 'necessary path'.

Waddington uses this landscape model to explain epigenetics, the role of changes in gene expression (rather than changes in genes), which alter the way in which an organism will develop. In another illustration (see Figure 4.12b), Waddington illustrates the underneath of the landscape, where there are 'guide ropes' controlled by genes (represented as pegs). The genes are fixed in a cell, but the gene expression is changeable; that is, the guide ropes can move, changing the landscape above. Because gene expression can be altered by environmental conditions, the epigenetic landscape changes in different ways. Changes in the landscape come about not necessarily because of the alteration of a gene (the stakes) but because of how the gene is expressed (the guide ropes).

Aside from the appealing visual nature of the explanation (this model has been quoted before in design contexts, for example, Longo and Zakhama, 2013), the epigenetic landscape model offers a useful way of considering biological fabrication for a number of reasons.

The inclined plan of the landscape represents the inevitability of change and transformation in biology, and if we imagine that our bacterial cells are on a journey from early development, we as designers want to influence that specialization by altering the landscape.

In biological fabrication, our role is to influence this landscape such that the ball (representing the development of the cell or set of cells) not only arrives at an end point which we determine but travels through a series of gullies representing assembly steps. We may choose to alter the landscape through various means, and, in synthetic biology, we are interested in the gene 'stakes', reconfiguring them to alter their expression and changing the landscape above. Some genes will have systemic effects, altering or even completely destroying the landscape above, others will have little or no effect and each will be dependent on the environment (Figure 4.13). Selective breeding offers another form of generic engineering, but, in this case, we are selecting for the most favourable landscape – altering its topography through generations (Figure 4.14) by altering the genes but without accessing them directly (as we would in genetic engineering).

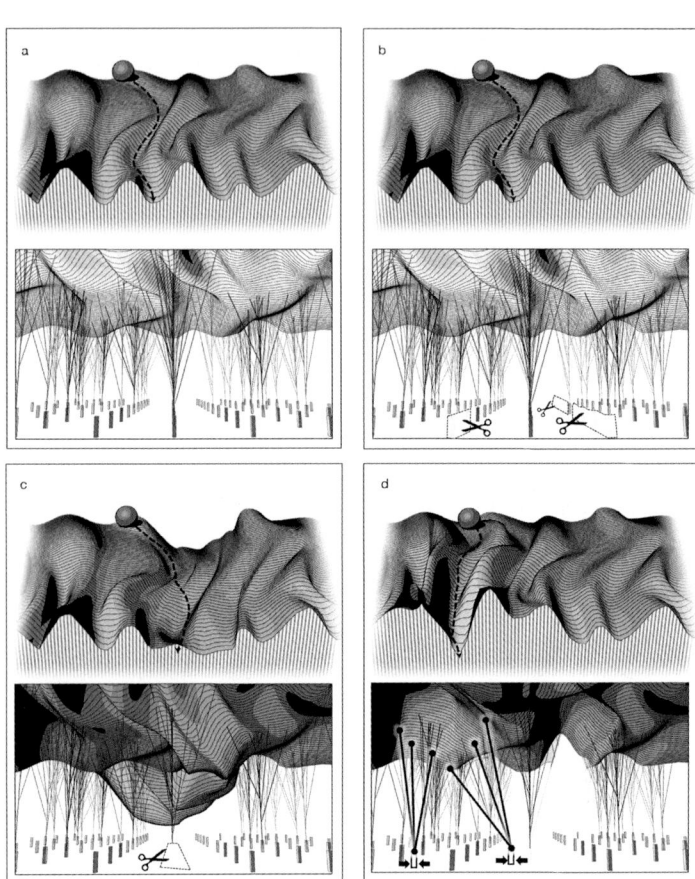

Figure 4.13 In synthetic biology, the developmental landscape is usually changed through the subtraction or addition of gene 'stakes'. The removal of single genes may have little or no effect (b), while the removal of some genes can cause a systemic altering of the cell's developmental destiny (c). New valleys can also be created through the addition of genes (d).

We also have access to controls above the landscape which are not represented in Waddington's diagram. What I have variously termed as 'top down' or 'designing from the outside' consists of ways in which we can alter the environment of a biological system to define our preferred outcomes. In the new diagram, Figure 4.15, I have shown a more complex surface. This time, the gene stakes and

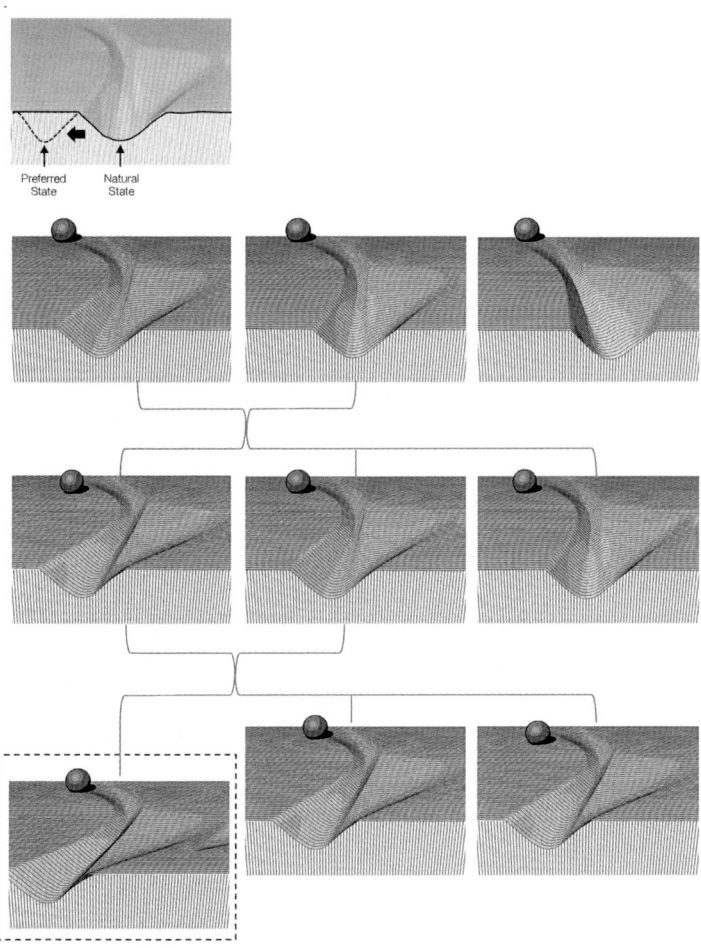

Figure 4.14 The developmental landscape can be changed over generations of selective breeding where genetic differences are exploited to enhance preferred traits.

guide ropes are presented as trees holding up the landscape above. I imagine them in compression rather than in tension. The fabric itself is not homogenous; in places it is rigid and resistant to change, and in other places it is flexible. In Figure 4.15b and c and Figure 4.15d and e, I have shown forces being applied to the surface. These forces

Fabrication in the living

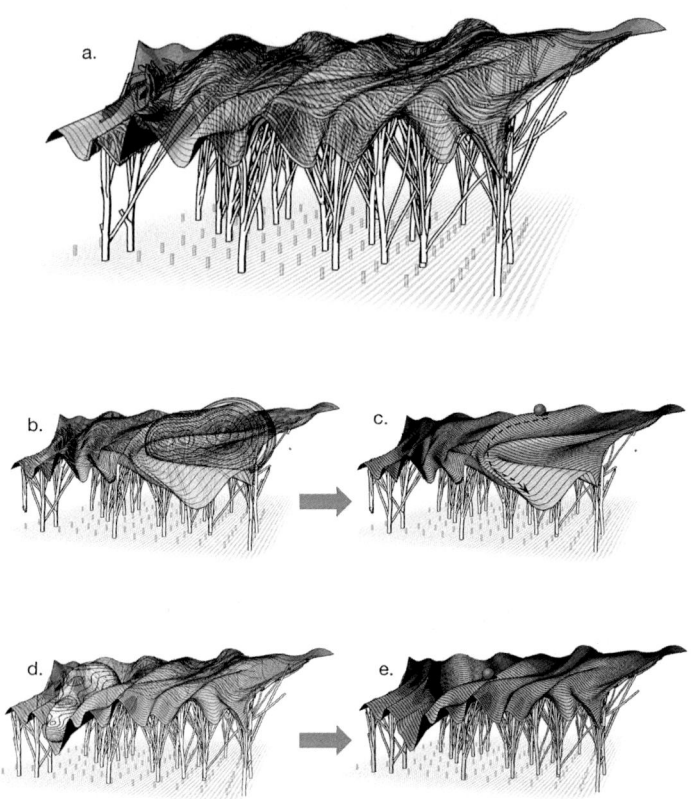

Figure 4.15 A revised version of Waddington's developmental landscape showing the effects of external influences on the 'creode' or necessary path of cell development.

represent general changes in the environment of the organism which produce more systemic changes in the surface of the developmental landscape. Because of the rigidities and flexibilities within the system and the interaction with the gene trees below, new pathways form as a result of the interaction between these forces and the gene expression structure.

COMPOSING THE MUSIC OF LIFE: CREODIC DESIGN
Drawing on the example of Waddington's epigenetic landscape, I am using the concept of a landscape of change and necessary

pathways as a metaphor to describe biological processes. Others have used alternative metaphors. Noble (2006), for example, uses music as his metaphor, making a case for the 'music of life', distinguishing between the information held in genes and the way in which genes orchestrate a process which includes the interaction of proteins, cells, organs and whole organisms in orchestral harmony. Noble is making a case for systems biology which looks at biological systems in a holistic way, often, argues Breitling, stemming from an aesthetic enjoyment of complexity and diversity, whilst seeking simplicity through general laws and principles (Breitling, 2010). It is perhaps this aesthetic enjoyment which leads to my (and others') distaste for the mechanistic models of biology proposed by fields such as synthetic biology. The challenge for us, however, is to prove that a system view of biology, used for analysis, can be turned into design frameworks which could lead to synthesis. Scientists often revel in complexity because complex problems always lead to more research papers, but in design, we need to find some level of simplicity and formal elegance.

Translating Waddington's idea of creodes into a design framework requires more than an alternative representation. The metaphor is far from complete. The landscape and the ball do not represent empirically defined states, and the diagram proper would be impossible to draw for a real cell, let alone multicellular systems. What I will describe here as creodic design, however, involves searching for the necessary paths for living cells to fabricate materials and for methods to alter the landscape towards a path which is favourable to the outcomes we want. This process, in contrast to Simon's distaste for the 'intuitive' and 'cookbooky' approach to designs he observed in *The Sciences of the Artificial*, involves something akin to a craft (Simon, 1996).

A good example of this is the work of Thora Arnadottir, who has been working on the process of biomineralization which I introduced previously. In her experiments, bacterial cells are seeded in a sand column, and a solution containing food for the bacteria and calcium source is pumped from the bottom of the tube upward. Over time, the bacteria grow in the sand and induce calcium carbonate to form – eventually cementing a column with a material with similar properties to concrete. While this approach is not novel, and others have produced this cemented material using similar methods (Dosier, 2011), it took more than a year of trial and error to perfect the technique. What works for a cylindrical column does not work for a cube, which we have found requires a 'vasculature' of tubes, inlets and outlets to cast the object. The samples which have gone wrong, however, are also revealing. This is not a straightforward casting process in which a liquid fills the vessel and then solidifies, taking on the form of the vessel. Even in the successful casts, the material is heterogeneous, exhibiting variability in texture and strength. The process of casting requires that the bacteria be able to survive and reproduce, and this includes access to oxygen, meaning that we see the highest levels of cementation at the top of the column. The chemical composition of the environment is also not homogenous, and the process of pumping the liquid through the column creates flows and voids. The material is, therefore, variably cemented, with dense material in some places and much looser and more friable material in others (Figure 4.16).

In the next set of experiments, Arnadottir will begin to explore how these variabilities in function and flow can be utilized, developing biological casting which makes use of these interactions between the living cells, the sand and the flows of media and oxygen in the system (Figure 4.17). These experiments are based on Arnadottir's growing intuition as to how the material will work and of the constraints and the capabilities of this method of casting or, as she describes it, the 'choreography' shaped by *in vivo* and *in vitro* processes. An understanding of processes at multiple scales is also required and, most importantly, of when to intervene. For her project, Arnadottir is developing a craft through what I would describe as a form of creodic design – defining the

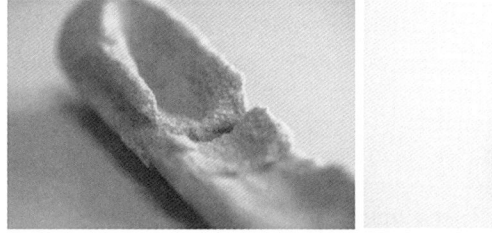

Figure 4.16 A bacteria-cemented column showing differential areas of cementation as the bacteria, interacting with the nutrients and calcium source, have selected selectively cemented sand grains.
Source: Image courtesy of Thora Arnadottir.

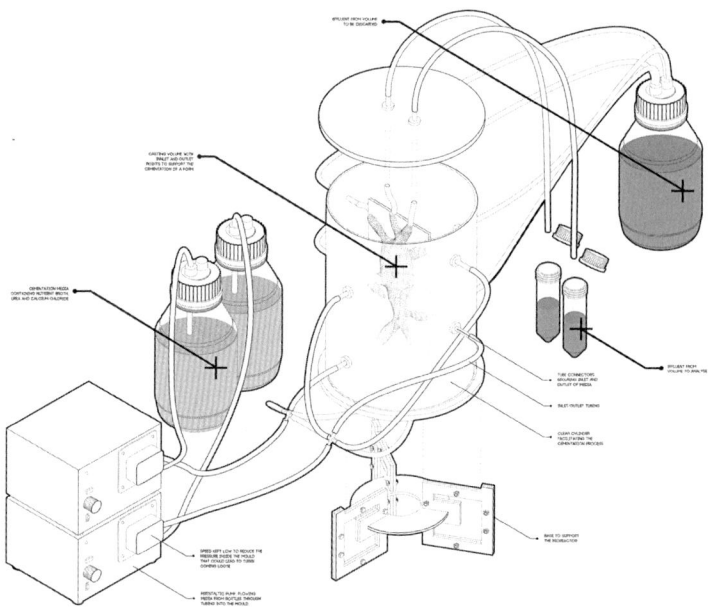

Figure 4.17 Arnadottir's experimental setup for the next phase of research.
Source: Image courtesy of Thora Arnadottir.

necessary spatial and temporal path for cells she is cultivating and for the development of her material to achieve a desired outcome.

CONCLUSION

In this chapter, I have attempted to move beyond the logics of assembly at individual scales described in Chapter 3 and have introduced alternative design frameworks which try to address the challenge of biological assembly at multiple scales simultaneously. I have argued that, while biological systems and biological fabrication can be analyzed in terms of a hierarchy of scales, the scales of assembly which, when combined, constitute biological fabrication are not discrete. There are, rather, causalities which cross multiple scales from both the top down and bottom up. I have also challenged the idea that biological systems can be thought of, in design terms, as being made up of decomposable structures. There are, I would suggest, no clear distinctions between inside and outside or separate decomposable parts. Rather there is a flow of energy and matter which can be disrupted and potentially choreographed to an end result.

I proposed that we should go beyond traditional engineering and design thinking to understand causality as a multidimensional landscape to be sculpted through genes and through environments, both *in vitro* and *in vivo*. To illustrate this, I have adapted Waddington's creode model, illustrated as pathways through landscapes of change. The model is far from complete. It does not offer an empirical basis for analysis or for synthesis but is, rather, an alternative to metaphors of hierarchies of scales and networks of parts. However, through the craft process developed by Arnadottir, I have hinted at how a creodic design process might be developed.

REFERENCES

Basu, S. *et al.* (2005) 'A Synthetic Multicellular System for Programmed Pattern Formation', *Nature*, 434(7037), pp. 1130–1134. https://doi.org/10.1038/nature03461.

Breitling, R. (2010) 'What Is Systems Biology?', *Frontiers in Physiology*, 1, pp. 1–5. https://doi.org/10.3389/fphys.2010.00009.

Cartwright, J. H. E. and Checa, A. G. (2007) 'The Dynamics of Nacre Self-Assembly', *Journal of the Royal Society, Interface/the Royal Society*, 4, pp. 491–504. https://doi.org/10.1098/rsif.2006.0188.

Chaudhuri, P., Bhattacharyya, S. and Samal, A. C. (2016) 'Living Root Bridge: A Potential No Cost Eco-Technology for Mitigating Rural Communication Problems', *International Journal of Experimental Research and Review*, 5, pp. 33–36.

Cosgrove, D. J. (2005) 'Growth of the Plant Cell Wall', *Nature Reviews Molecular Cell Biology*, 6(11), pp. 850–861. https://doi.org/10.1038/nrm1746.

Dade-Robertson, M. *et al.* (2017) 'Architects of Nature: Growing Buildings with Bacterial Biofilms', *Microbial Biotechnology*, 10(5). https://doi.org/10.1111/1751-7915.12833.

Dade-Robertson, M. *et al.* (2016) 'Thinking Soils: A Synthetic Biology Approach to Material Based Design Computation', in *ACADIA*. Ann Arbor: University of Michigan, pp. 460–469.

Dade-Robertson, M., Ramirez-Figueroa, C. and Zhang, M. (2015) 'Material Ecologies for Synthetic Biology: Biomineralization and the State Space of Design', *Computer-Aided Design*, 60, pp. 28–39.

Decho, A. W. (2010) 'Overview of Biopolymer-induced Mineralization: What Goes on in Biofilms?', *Ecological Engineering*, 36(2), pp. 137–144. https://doi.org/10.1016/j.ecoleng.2009.01.003.

Dosier, G. (2011) 'Methods for Making Construction Material Using Enzyme Producing Bacteria', US Patent 20110262640 A1.

Gilbert, S. F. et al. (2000) 'Embracing Complexity: Organicism for the 21st Century', *Developmental Dynamics*, 219, pp. 1–9.

Guyet, A. et al. (2018) 'Mild Hydrostatic Pressure Triggers Oxidative Responses in Escherichia coli', *PLOS ONE*, 13(7), pp. 1–19. https://doi.org/10.1371/journal.pone.0200660.

Henriques, G. (2003) 'The Tree of Knowledge System and the Theoretical Unification of Psychology', *Review of General Psychology*, 7(2), pp. 150–182. https://doi.org/10.1037/1089-2680.7.2.150.

Ishii, A. et al. (2005) 'Analysis of Hydrostatic Pressure Effects on Transcription in Escherichia coli by DNA Microarray Procedure', *Extremophiles*, 9(1), pp. 65–73. https://doi.org/10.1007/s00792-004-0414-3.

Jonkers, H. (2007) 'Self Healing Concrete: A Biological Approach', *Self Healing Materials*, 100, pp. 195–204. https://doi.org/10.1007/978-1-4020-6250-6_9.

Korner, C., Pelaez, M.-R. S. and John, P. (1989) 'Why Are Bonsai Plants Small? A Consideration of Cell Size', *Australian Journal of Plant Physiology*, 16, pp. 443–448. https://doi.org/10.1071/pp9890443.

Lin, A. Y.-M., Chen, P.-Y. and Meyers, M. A. (2008) 'The Growth of Nacre in the Abalone Shell', *Acta Biomaterialia*, 4(1), pp. 131–138. https://doi.org/10.1016/j.actbio.2007.05.005.

Longo, G. and Zakhama, N. (2013) 'The Model as an Organizing Outlook upon the Real', in *Naturalizing Architecture*. Orleans: HYX, pp. 158–187.

Neidhardt, F., Ingraham, J. and Schaechter, M. (1990) *Physiology of a Bacteria Cell: A Molecular Approach*. Sunderland, MA: Sinauer Associates.

Nguyen, P. Q. et al. (2018) 'Engineered Living Materials: Prospects and Challenges for Using Biological Systems to Direct the Assembly of Smart Materials', *Advanced Materials*, 30(19), pp. 1–34. https://doi.org/10.1002/adma.201704847.

Noble, D. (2006) *The Music of Life: Biology Beyond Genes*. Oxford: Oxford University Press.

Ouzounis, C. A. and Karp, P. D. (2000) 'Global Properties of the Metabolic Map of *Escherichia coli*', *Genome Research*, 10(4), pp. 568–576. https://doi.org/10.1101/gr.10.4.568.

Roberts, M. A. J. et al. (2013) 'Synthetic Biology: Biology by Design', *Microbiology (United Kingdom)*, 159(PART7), pp. 1219–1220. https://doi.org/10.1099/mic.0.069724–0.

Ross, P., Mayer, R. and Benziman, M. (1991) 'Cellulose Biosynthesis and Function in Bacteria', 55(1), pp. 35–58.

Sato, T. et al. (1996) 'High Pressure Represses Expression of the malB Operon in *Escherichia coli*', *FEMS Microbiology Letters*, 135(1), pp. 111–116. https://doi.org/10.1016/0378–1097(95)00438–6.

Shoda, M. and Sugano, Y. (2005) 'Recent Advances in Bacterial Cellulose Production', *Biotechnology and Bioprocess Engineering*, 10(1), pp. 1–8. https://doi.org/10.1007/bf02931175.

Simon, H. A. (1996) *The Sciences of the Artificial*. Cambridge, MA: MIT Press.

Stano, P. and Mavelli, F. (2015) 'Protocells Models in Origin of Life and Synthetic Biology', *Life*, 5(4), pp. 1700–1702. https://doi.org/10.3390/life5041700.

Stotz, K. and Griffiths, P. E. (2017) 'Biological Information, Causality and Specificity', in Walker, S., Davies, P. and Ellis, G. (eds) *From Matter to Life: Information and Causality*. Cambridge: Cambridge University Press, pp. 366–390.

Vincent, J. (2012) *Structural Biomaterials*. 3rd edn. Princeton: Princeton University Press.

Waddington, C. H. (2014) *The Strategy of Genes: A Discussion of Some Aspects of Theoretical Biology*. Abingdon: Routledge.

Zolotovsky, K. (2017) *Guided Growth: Design and Computation of Biologically Active Materials*. Cambridge, MA: MIT Press. https://doi.org/10.1097/BPO.0000000000001022.

Zolotovsky, K., Gazit, M. and Oritz, C. (2017) 'Guided Growth: The Interplay among Life, Material and Scaffolding', in Tibbits, S. (ed.) *Active Matter*. Cambridge, MA: MIT Press, pp. 83–88.

Zolotovsky, K., Gazit, M. and Oritz, C. (2018) 'Guided Growth of Bacterial Cellulose Biofilms', in Vasiliki, V. *et al.* (eds) *Biomimetic and Biohybrid Systems*. Cham: Springer International Publishing, pp. 538–548.

Conclusion: the craft of living construction

Chapter 5

THE END OF THE BEGINNING

Making a bacterial cell grow and produce a material in useful quantities in a lab is not an easy feat. I have often given conference presentations alongside colleagues in other architecture research groups where my images of crystals shown in electron microscopes has needed to compete with images of pavilions and even whole buildings. Working with living cells is not like working with other media. You can't master life in the way a painter masters oils or a joiner masters wood. So, where do we start? This book is designed to be one potential staring point.

In Chapter 2, I set up a dichotomy between the natural and the artificial. The contemporary discourse which sees biotechnology as an emerging technology frequently makes reference to the blurring of the lines between the natural and the artificial, but this position requires that we be able to clearly define the natural and the artificial in the first place and that this 'blurring' is a contemporary issue. In reality, very little in the biological world now exists without human intervention, and years of selective breeding have created new and strange species from dogs and cats to roses, which exist because of human preferences and 'design' rather than evolutionary necessity. These species are adapted to the human environment rather than the 'natural' environment more broadly conceived. What has changed is our tools for manipulation. Modern techniques from molecular biology allow us to make precise changes in the DNA of living cells, taking out the need for random mutation and selection used in selective breeding.

There is a distinction in such practices between an 'outside' and an 'inside' of a biological system where the inside is delineated by the membranes of cells. The sciences of molecular biology and, more recently, synthetic biology provide tools which allow us to change the *inside* of a biological system, in contrast to processes such as selective breeding and farming which allow us only to modify the *outside*. This inside/outside distinction also allows for a conversation which places biological systems in the realm of artificial things. Using Simons' *Sciences of the Artificial*, I showed that his definition of artificial relied on the separation of an artificial inside (using the example of the workings of a ships chronometer) and a natural outside (the waves). In order to work, artificial things must be insulated from natural things other than through designed interfaces. 'Designed' objects, therefore, are always designed on the inside, and we never design their outside.

Bio Design, however, must of necessity work with a system which is already made. We can't (yet) construct a novel living cell from scratch, and cells are not robots. Even in a lab environment, much of the control of biological cells comes not from techniques such as genetic manipulation but through the control of the environment, that is, control from the outside.

Anyone who has worked in a microbiology lab will know that great effort goes into ensuring that cells are grown at specific temperatures with constrained chemicals and conditions. Take away the lab coats, and microbiology looks more like indoor micro-agriculture. This tight control of the environment is the case because, in reductive science, we want to reduce complexity – to know that a single change, for example, in the DNA of bacteria, is going to lead to a clear and repeatable result and not be affected by a plethora of other influences. This approach to complexity clearly has limitations, however. If we want to take biotechnologies out of the lab and into the wild, we cannot rely on strictly bounded environmental conditions – they will be unpredictable. Because our system is already mostly made, it will behave in ways which we might not have anticipated.

In Chapter 3, I began to break down this inside and outside model by looking at the idea of assembly in biological systems in terms of multiple scales through the patterning of energy and matter over time and asking: Where is the information for the assembly of matter? Designing the assembly of biological systems is a matter of intervening to change the information content of a system and, depending on scale, we find the information in different parts of the system. There are also different logics of assembly, and these lead to different means of interventions and different scientific disciplines and tools. As we stack these scales one on top of another, we also see emergent properties (what I described as information for free) as simple parts become complex wholes.

In seeking to understand complexity, we tend to seek out hierarchies and to define bounded 'decomposable' niches, where individual parts of a system are (mostly) separate from the others. These decomposable systems give those who wish to design using biology hope that, like more traditionally engineered systems, biological systems can be viewed as a series of separated discrete and potentially interchangeable units. Framing biological systems in terms of a hierarchy also gives rise to ideas of top-down v. bottom-up processes in design, with a claim that these biological systems operate primarily through bottom-up processes of assembly.

In Chapter 4, I began to examine biological fabrication, giving an example of projects in which we see causalities across scales both from the bottom up and from the top down. While simple in their scope, these projects show that there is an interaction of many forces in biological fabrication and that we can distinguish between the scale of the intervention and the scales of its effects. As an alternative model, I proposed one where domains of information, *in vivo*, *in vitro* and *in silico*, are seen as interacting and where projects involving living fabrication can be mapped, depending on the use of tools and techniques associated with each domain.

Using Waddington's landscape of development, I also proposed a model which shows the flow of energy and matter and the tensions and trajectories of fabrication in biology. I proposed that this could be part of an alternative model that understands this landscape as a membrane with rigidity and flexibility in different places and which can be affected from 'above and below'. This radical abstraction is far from a model of design but, from my perspective, offers some insight which might lead to design directions and an alternative to the machine metaphor of biology. Understating this landscape seems vital in understanding how we can design with biological systems.

CRAFTING BIOLOGY

Imagine an experienced glass blower creating a new vase. In tackling the new project, they have to deal with an extraordinary number of variables and processes which operate at multiple scales. They have to understand the molecular structure of the glass, the way in which the silica molecules will stabilize to form the stiff but brittle final material and how, through the careful use of additives throughout the process, they can achieve a range of different material outcomes, including the colour, transparency and finish of the material. They must be able to work the material in its molten state, making fine judgements about when the material is liquid enough to work but not so molten as to fall off their blowpipe. They need to anticipate the thermal dynamics of the material as it clings to the end of their blowpipe and rapidly cools, differences in cooling between the inner surface of the glass and its outer surface, the hottest location within the furnace and the interaction between their breath and the material as they shape the inner void of the vase. They must understand the effect of physics on the glass as gravity pulls the molten material to the ground, constantly rotating the blowpipe to achieve an even thickness and form. Taken further, the glass blower who wishes to experiment with this process, creating something truly new, must understand the parameters of the material they are using whilst pushing their process and techniques to the limits. The process is carefully choreographed and, to the outside viewer, the subtle embodied decisions made by the glassblower may be imperceptible.

We call this process 'craft', and I would suggest it offers some insight into a future for living construction. The intuition exhibited by the experienced glass blower is, perhaps, impossible to achieve in biology. Materials, the glass blower example reminds us, have agency, but the degree of agency is radically increased when we deal with living cells. The glass blower deals with a particular intersection of physics and chemistry, of thermodynamics and their own physical choreography. Yet they don't need to understand their system in its totality. What I describe here as intuition is, in more precise terms, an understanding of cause and effect without necessarily understating the process in between. In biology, of course, there are many more processes in between. The interactions between the dynamics inside a cell and its interactions with the environment are far from straightforward and change over time. This complex choreography may be, for the time being, beyond our full understanding. However, in my extension of Waddington's epigenetic landscape model, I have proposed that to understand the landscape of biological construction is to understand and be able to articulate the tensions and flexibilities within the system and the intersection of different processes inside and outside a cell. To map this territory might not require a complete understanding, but it does require that we find the appropriate mapping techniques and not be bounded by particular scales and the tools they are associated with.

In architecture, one of the most often cited examples of biological fabrication is the use of mycelium, the root network of fungus, as a material. Mycelium can grow fast on a range of waste materials and act as a binder, creating material of significant strength and insulative properties. Unlike the examples of bacterial cellulose and

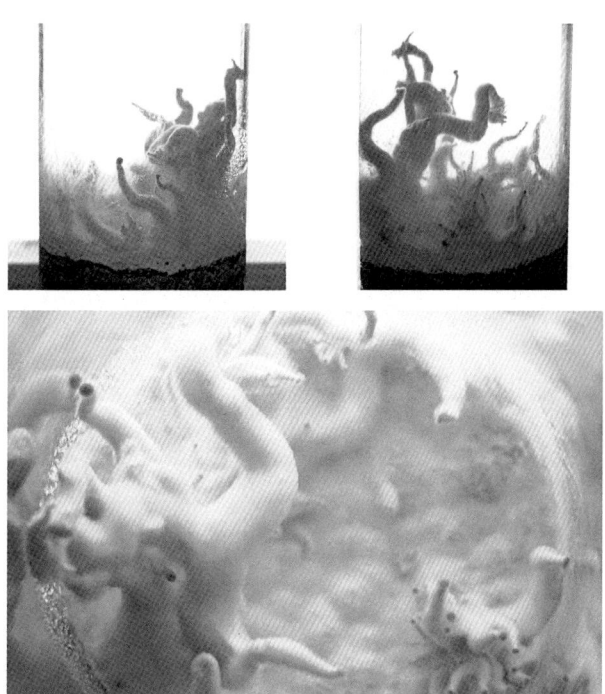

Figure 5.1 Mushroom growth constrained within a glass container.
Source: Image courtesy of Dilan Ozkan.

biomineralization, the materials and structures grown with mycelium are easily scalable, and pioneering projects by, for example, the architectural practice The Living, demonstrate a large-scale pavilion made substantially out of mycelium-based bricks. As important and pioneering as the Hy-Fi pavilion is, I sometimes use the project to pose a question to my research team. In an echo of Louis Kahn's 'I say to a brick, "What do you want brick?"', we ask whether mycelium wants to be a brick.

The truth is that mycelium probably doesn't care much about what it is so long as it can survive and reproduce, although mycelium-based materials tend to be produced on the basis that fruiting bodies, that is, mushrooms, are not allowed to form. My question raises a point about whether the results produced by the relationship between aggregate and mould in the production of clay bricks are the same for mycelium bricks. In trying to understand our materials and the shape of our creode for the fabrication of materials, we need to pay special attention to the relationship between the growth and development of the organisms we work with and their complex relationship to the environment we are creating for them.

In her experiments, Dilan Ozkan has been asking the question of what mycelium wants, entering into conversations with the growing mushrooms by intervening in their growth. In the photographs she takes (Figure 5.1), it is tempting to see the physical manifestations of creodes in the undulations of the surface skin of the forming mushrooms. It would be dangerous to take this observation too literally – a creode is metaphor standing in for an abstract data space – however, we do begin to sense, in the confines of Ozkan's glass jars, the interface between the inside and outside of the biological system and its *in vivo* dialogue with the *in vitro* container.

A NEW AGRICULTURAL REVOLUTION

It is often said that we are entering a fourth industrial revolution through the merging of the physical, digital and biological materials and systems. However, when I observe the work of emerging bio designers, hear the replacement of words like 'production' for 'cultivation' and note how much time is spent in the lab nurturing living cells, I wonder whether we are entering a new agricultural revolution (the fourth or fifth, depending on your perspective).

This form of agriculture does not necessarily take place in the typical environments of agricultural production, but may take place in labs or not-yet-invented fabrication facilities. It will be about design from the outside as well as from the inside.

It will not necessarily result in objects and materials which look 'biological', but rather the shapes of which result from their patterns of construction and the interaction of many multiscale parameters.

It will be much like the old types of agriculture. It will be a craft process where complexity is managed by developing intuitions on how living cells grow, propagate and synthesize materials in relation to their inner logics and outer forces.

However, it will result in a greater range of materials and products.

This new type of agriculture will make use of molecular knowledge, cultivating molecular interaction as much as growth, but it will require new ways of visualizing and modelling such systems, offering new design intuitions and landscapes of change.

Index

Note: Page numbers in *italics* indicate a figure on the corresponding page.

agriculture 25, 96
Alexander, C. 76
alien technology 33
amino acids 32, 38–39, 51, 58
Aristotle 21, 23
Armstrong, R. 3, 6, 52
Arnadottir, T. 87, 88, *89*
artificial life 6, 60
artificial technology 13
artificial things: biological systems in the realm of 95; boundaries for the study of 16; defined 17; synthesis 20; vs. natural things 21
assembly: chemical processes 35; distinction from fabrication 32; molecular 35, *36*; multicellular 42; of non-organic materials 40; templated *39*, 40; *see also* biological assembly
avidity 39

bacteria: as compared to robots 14; gene expression 69; sensor and builder 71
Bensaude-Vincent, B.: on distinguishing artificial and natural products 21; on machine metaphors 14
Bio Design (Biological Design) 3, 26, 95
Biodesign Challenge 5
biological assembly 34, 40, 76
biological materials 5, 6
biological systems 3, 18, 19; inside and outside of 16, 77, 89, 95, 96; machine metaphors 14; use of term 6
bottom-up causality 67, 97
bottom-up design 60, 61, 63
Boudry, M. 24
Built to Grow 6

Calvert, J. 21
canalization 81
Catts, O. 24
cell 6; acting as templates 40; assembly information held by 43; bacterial cells 14; bacterial cellulose 63; cellulose 61; fabrication and assembly 32; inside and outside 25; intracellular chemistry 36; moulding 46; subject to mechanical forces 45
chemical processes 35
chemotaxis *44*
Cogdell, C. 2
creodic design 87

Davies, J. 33, 43, 46
decomposability 76, 77, 96
deoxyribonucleic acid (DNA) 3, 34
design 16, 17
design speculation 2–3
diffusion limited aggregation (DLA) 48, 49

emergence 47, 50, 53, 57, *58*
"End of the Artificial" 7, 13
energy 35, 36

engineered living materials (ELMs) 61
epigenetic landscape 83, 86

fabrication 32, 58, 60–62
forcible constraints 21, 22, 25

gene regulation 18
Gilbert, S. F. 60
Guided Growth project 68

Henriques, G. 57, *58*
Hensel, M. 2
heterodimers 39
Hillis, D. 13
homomers 39

information 33, 35
information for free 47, 50, 57
in silico 67, 78, 97
International Competition for Genetically Engineered Machines (iGEM) 1, 5
intracellular chemistry 36
intracellular interaction 43
in vitro and *in vivo* 57, 62, 63, 78, 80, 88, 90, 97
Ishii, A. 69
Ito, J. 13

Johnson, M. 26
Journal of Design and Science (*JoDS*) 13

Kahn, L. 99
kludging 24, 25

Lakoff, G. 26
landscape of development 97

Lazebnik, Y. 21
Lee, K. 22
life 6
living architectures 2, 6
Living Architecture Systems Group (LASG) 6

machine metaphor 13, 14, 15, 18, 23, 24, 25, 26; *see also* organisms vs. machines
McLain, S. 34
Menges, A. 3
M. genitalium see Mycoplasma genitalium
MIT Bio Summit 5
morphogenesis 43, 44
multicellular assembly 43
Mycoplasma genitalium 20, 33
mycelium 98–99

nature's own agencies 23, 24, 25, 26
Newman, W. R. 15, 21
Nguyen, P. Q. 61
Nicholson, D. 24
Noble, D. 87
Notes on the Synthesis of Form (Alexander) 76

organisms vs. machines 14, 24
Oxman, N. 3
Ozkan, D. 100

"Physics of Life, The" (McLain) 34
Pigliucci, M. 24
plastics 5
Preston, B. 23

Prigogine, I. 52
protein folding problem 38
proteins 38, 40, 51, 68
protocells 6

Rittel, H. 16
RNA sequencing 69
Royal Institution, The 34

Sato, T. 69
Schon, D. 16
Sciences of the Artificial, The (Simon) 7, 15, 16, 17, 23, 87
self-assembly 33, 50, 51
Simon, H. 7, 15, 16, 17, 19, 20, 21, 22, 23, 25, 87, 95
Strategy of Genes, The (Waddington) 80
synthetic biology 2, 3, 7, 13, 21, 23, 60, 77, *79*, 95; contemporary 17–18, 25; definition 18; parts, devices, and systems 17; practice of 21–24

technology readiness levels *8*
Thinking Soils project 67–68
Tibbits, S. 3, 41
top-down causality 73, 76, 78, 97; *see also* bottom-up causality
transcription 40
translation 40
tree of knowledge *58*
Turing patterns 47

Venter, C. 33
vertical causalities 67
Vincent, J. 58

Waddington, C. H. 80, 81, 86

Zhang, S. 51
Zolotovsky, K. 64–68
Zurr, I. 24